Transforming Human Resource Functions With Automation

Anchal Pathak
Bule Hora University, Ethiopia

Shikha Rana
IMS Unison University, India

A volume in the Advances in
Human Resources Management
and Organizational Development
(AHRMOD) Book Series

Published in the United States of America by
  IGI Global
  Business Science Reference (an imprint of IGI Global)
  701 E. Chocolate Avenue
  Hershey PA, USA 17033
  Tel: 717-533-8845
  Fax: 717-533-8661
  E-mail: cust@igi-global.com
  Web site: http://www.igi-global.com

Library of Congress Cataloging-in-Publication Data

Names: Pathak, Anchal, 1985- editor. | Rana, Shikha, 1984- editor.
Title: Transforming human resource functions with automation / Anchal
  Pathak and Shikha Rana, editors.
Description: Hershey, PA : Business Science Reference, [2020] | Includes
  bibliographical references and index. | Summary: "This book is a
  collection of research on the methods and applications of artificial
  intelligence and autonomous systems within human resource management and
  modern alterations that are occurring"-- Provided by publisher.
Identifiers: LCCN 2020004194 (print) | LCCN 2020004195 (ebook) | ISBN
  9781799841807 (hardcover) | ISBN 9781799854036 (paperback) | ISBN
  9781799841814 (ebook)
Subjects: LCSH: Personnel management--Technological innovations. | Manpower
  planning.
Classification: LCC HF5549.5.T33 T73 2020 (print) | LCC HF5549.5.T33
  (ebook) | DDC 658.300285--dc23
LC record available at https://lccn.loc.gov/2020004194
LC ebook record available at https://lccn.loc.gov/2020004195

This book is published in the IGI Global book series Advances in Human Resources Management
and Organizational Development (AHRMOD) (ISSN: 2327-3372; eISSN: 2327-3380)

British Cataloguing in Publication Data
A Cataloguing in Publication record for this book is available from the British Library.

All work contributed to this book is new, previously-unpublished material.
The views expressed in this book are those of the authors, but not necessarily of the publisher.

For electronic access to this publication, please contact: eresources@igi-global.com.

Advances in Human Resources Management and Organizational Development (AHRMOD) Book Series

ISSN:2327-3372
EISSN:2327-3380

Editor-in-Chief: Patricia Ordóñez de Pablos, Universidad de Oviedo, Spain

MISSION

A solid foundation is essential to the development and success of any organization and can be accomplished through the effective and careful management of an organization's human capital. Research in human resources management and organizational development is necessary in providing business leaders with the tools and methodologies which will assist in the development and maintenance of their organizational structure.

The **Advances in Human Resources Management and Organizational Development (AHRMOD) Book Series** aims to publish the latest research on all aspects of human resources as well as the latest methodologies, tools, and theories regarding organizational development and sustainability. The **AHRMOD Book Series** intends to provide business professionals, managers, researchers, and students with the necessary resources to effectively develop and implement organizational strategies.

COVERAGE

- Talent Identification and Management
- Skills Management
- Employment and Labor Laws
- Organizational Development
- Personnel Policies
- Employee Evaluation
- Human Resources Development
- Performance Improvement
- Executive Education
- Coaching and Mentoring

IGI Global is currently accepting manuscripts for publication within this series. To submit a proposal for a volume in this series, please contact our Acquisition Editors at Acquisitions@igi-global.com or visit: http://www.igi-global.com/publish/.

The Advances in Human Resources Management and Organizational Development (AHRMOD) Book Series (ISSN 2327-3372) is published by IGI Global, 701 E. Chocolate Avenue, Hershey, PA 17033-1240, USA, www.igi-global.com. This series is composed of titles available for purchase individually; each title is edited to be contextually exclusive from any other title within the series. For pricing and ordering information please visit http://www.igi-global.com/book-series/advances-human-resources-management-organizational/73670. Postmaster: Send all address changes to above address. Copyright © 2021 IGI Global. All rights, including translation in other languages reserved by the publisher. No part of this series may be reproduced or used in any form or by any means – graphics, electronic, or mechanical, including photocopying, recording, taping, or information and retrieval systems – without written permission from the publisher, except for non commercial, educational use, including classroom teaching purposes. The views expressed in this series are those of the authors, but not necessarily of IGI Global.

Titles in this Series

For a list of additional titles in this series, please visit:
http://www.igi-global.com/book-series/advances-human-resources-management-organizational/73670

Future of Work, Work-Family Satisfaction, and Employee Well-Being in the Fourth Industrial Revolution
Ethel Ndidiamaka Abe (University of KwaZulu-Natal, South Africa)
Business Science Reference • © 2021 • 382pp • H/C (ISBN: 9781799833475) • US $215.00

Handbook of Research on Cyberbullying and Online Harassment in the Workplace
Leslie Ramos Salazar (West Texas A&M University, USA)
Business Science Reference • © 2021 • 717pp • H/C (ISBN: 9781799849124) • US $265.00

Integration and Application of Business Graduate and Business Leader Competency-Models
Donta S. Harper (University of Washington, USA)
Business Science Reference • © 2021 • 275pp • H/C (ISBN: 9781799865377) • US $195.00

Human Resource Management Practices for Promoting Sustainability
Sulaiman Olusegun Atiku (Naimbia University of Science and Technology, Namibia) and Tinuke Fapohunda (Lagos State University, Nigeria)
Business Science Reference • © 2021 • 326pp • H/C (ISBN: 9781799845225) • US $195.00

Effective Strategies for Communicating Insights in Business
Ross Jackson (Wittenberg University, USA) and Amanda Reboulet (Air Force Civilian Service, USA)
Business Science Reference • © 2021 • 330pp • H/C (ISBN: 9781799839644) • US $195.00

Anywhere Working and the Future of Work
Yvette Blount (Macquarie University, Australia) and Marianne Gloet (University of Melbourne, Australia)
Business Science Reference • © 2021 • 287pp • H/C (ISBN: 9781799841593) • US $195.00

For an entire list of titles in this series, please visit:
http://www.igi-global.com/book-series/advances-human-resources-management-organizational/73670

701 East Chocolate Avenue, Hershey, PA 17033, USA
Tel: 717-533-8845 x100 • Fax: 717-533-8661
E-Mail: cust@igi-global.com • www.igi-global.com

Table of Contents

Detailed Table of Contents

Raluca Bunduchi, University of Edinburgh, UK
Aizhan Tursunbayeva, University of Molise, Italy
Claudia Pagliari, University of Edinburgh, UK

Previous research demonstrates that actors seek acceptance of innovations during the diffusion and adoption process, but exactly what kind of acceptance they seek and how they go about obtaining this remains unclear. This study interrogated academic databases and Twitter to identify and analyze research and grey literature concerning blockchain adoption in human resources (HR). The data was analyzed through the theoretical lens of pragmatic, moral, and cognitive legitimacy to explore how potential adopters of this innovation conceive its value and alignment with their goals. The results show how pragmatic considerations are emphasized around streamlining current HR processes. Arguments for moral legitimacy are considered far less, while narratives on cognitive legitimacy are absent. The findings also show high levels of conformity around legitimation strategies within the community, with only some evidence of selection and manipulation strategies. The authors discuss the implications of these findings for the adoption and diffusion of blockchain in the HR sector.

Preeti Bhaskar, ICFAI University, Dehradun, India & University of Technology and Applied Sciences, Ibra, Oman
Muddu Vinay, ICFAI University, Dehradun, India
Amit Joshi, ICFAI University, Dehradun, India

The United Nations have sustenance e-government to promote good governance for achieving Sustainable Development Goals – 2030 by fostering environmental, technological, socio-cultural, and political sustainability. Many countries are making sincere efforts to create an ecosystem for fostering e-government to contribute to Sustainable Development Goals. The government is investing a huge amount in e-government projects, but most of them are not able to meet the desired results. It has been reported that most of the e-government is failing due to the non-adoption behavior of employees. This chapter seeks to understand the employee's perspectives by discussing benefits, barriers, and enablers to adopting e-government through a systematic literature review.

Chapter 3

Innovation has become an integral part of every business organization because it provides sustainable competitive advantage to the company. In today's highly dynamic business environment every organization wants to succeed by leveraging their employees' talent. In order to leverage the employees' talent, it is important to follow relevant HRM policies or to continuously introduce innovative HR practices to meet the expectations of the employees. Reviewing the literature and previous research work, this study has tried to find out the various innovative HR practices initiated by Indian organizations till now. The findings will help in guiding how much more innovative practices are still to be initiated in order to attain employee confidence and loyalty for the company. The study is beneficial for business leaders, students, practitioners, and researchers.

Chapter 4

The new age world, which has become part of our lives, is a world of rapid innovations and changing technologies. These changes bring new opportunities for organizations to exchange information, news, ideas, and work. Attracting and retaining best of employees has become the most crucial tactical problems for the people's department of companies all over the world. In the current scenario of high competition, the internet has substantially converted the features of recruitment and selection procedure of the businesses. The chapter presents an exploratory research on the impact of social media on recruitment in new age organizations.

Chapter 5

Over the last few years, the way of talent acquisition has evolved in different forms from attracting personal applications to getting connected with talented candidates through social networking sites. Recruitment through social networking platforms is putting a significant contribution in analyzing and hiring the right and best talent for an opening, and companies can't just ignore the potential and influence of these media platforms. These social platforms connect companies to potential hires and increase visibility by getting them connected to a huge audience. The future of recruitment lies in social media and companies cannot just ignore their presence due to prevailing challenges. It is important to find out viable solutions to the challenges organizations facing while using social media platforms in talent acquisition. The focus of this chapter is to capture strategies mitigating these challenges and suggest probable and profitable suggestions to companies for better utilization of social networking sites for effective recruitment.

HR is evolving into a more technology-based profession because organizations needs to streamline HR processes and reduce administrative burden, reducing administrative cost; compete more effectively with global talent; improve services and access data to the employees and managers; provide real-time metrics in order tom on spot decisions for the decision makers; and manage the workforce more effectively and enable the HR to transform so it can play more strategic role in the business and operations. The purpose of this chapter is to develop a meaningful debate on the innovations in human resource in terms of new ideas, methods, and technology to better meet the evolving requirement of the organization and workforce. Anticipating and exploring the future needs and circumstances rather than simply finding some responses to the situation, this chapter highlights challenges and prospects related to innovations in HR.

The chapter aims to investigate the influence of information technology, trust, rewards, leadership, and organizational culture on the knowledge sharing behavior of the employees that ultimately drives employee creativity. Drawing from the literature on employee creativity, knowledge sharing, and its influencing variables, this paper proposed a model comprising all such prominent variables and tested it quantitatively. For this purpose, 405 questionnaires were collected at Indore, India, and structural equation modeling was used to test the hypotheses. The findings show that organizational culture followed by leadership was the prominent factor affecting the knowledge sharing behavior of employees. Information technology, trust, and rewards followed next, respectively. Employee creativity was found to be significantly affected by knowledge sharing behavior. The study augments the research on employee creativity and knowledge sharing.

Chapter 8

This chapter presents the relationship and impact of employee empowerment on organizational trust. The purpose of this study is to analyze the relationship between employee empowerment and organizational trust in the IT industry. This study examines the difference between the empowerment level and organizational level of male and female employees. The questionnaire employed in this study consisted of empowerment by A. K. Mishra and G. M. Spreitzer and organizational trust variables by P. Mishra. The sample for the study consisted of 475 IT professionals from five IT organizations. Simple random sampling was used as a sampling technique, and this study was an ex-post in nature. Data were analyzed using t-test, correlation, and multiple regression. The result revealed that employee empowerment had a positive and significant impact on organizational trust. There is no significant difference in the empowerment of IT industry employees. It means that both male and female employees were equally empowered in their jobs. Male employees are more trust in their jobs as compared to female employees.

Chapter 9

Human resource (HR) management is all about people; there is no doubt about it. However, in the contemporary era inbuilt by the technological revolution and transformative development, the tenets of HR are finding a new footing. Technology applications have changed facets of corporate houses from restructuring organizations to resources to manpower. It has been seen that HR technologies are playing a major role in managing talents in organizations. Artificial intelligence, robotic process automation, and machine learning add an edge to talent acquisition and management. Machine learning tools are primarily being used in acquiring talents and enabling the hiring process effectively and efficiently. Today, as every company is moving to a new era of digitalization and data management, managing and mapping talents pose a challenge to C-suite and board-level management. The present study highlights the role of technology in managing talents in a series of HR processes.

Preface

Artificial Intelligence and robotics have grown widely in recent era. Automation refers to the practice of power-driven or computerized processes to accomplish work without—or with reduced—interference by humans. Examples include robots that flip hamburgers, computer algorithms that eliminate human employees in medical and legal offices, and driverless automobiles and aerial drones. According to Ravin Jesuthasan in his study on *Reinventing Jobs: A Four-Step Approach for Applying Automation to Work* (Harvard Business Review Press, 2017) automation drives advancement not by eliminating jobs but by eliminating particular job functions at which humans are inefficient, inconsistent or exposed to risk.

Today we have become highly dependent on technology, computer and rise of robotics has created a huge impact on the organization and varied functions of human resource management. Scientists are creating systems which are taking the place of human mind at present and future. With the advancement in technology and computer intelligence futuristic leadership and managing employees in organizations has become a huge challenge. Robotic process automation is a process by which unexciting and boring jobs, such as data entry and creating spreadsheets, are executed by a robot rather than a human. It has created more efficiency at the job which could possibly replace future human resource requirement at the global level which is the utmost matter of concern.

Researchers at Oxford University stated in their report that about 47% of all jobs could be completely overtaken by robots within the next 17 years. Similarly, the Gartner Group published outcomes that believe on a global scale, a third of our jobs will be lost to robots by 2025. There are many other researchers who have same findings for the upcoming years. In this future, one of the biggest issues which will be faced by HR departments is that they have to find out which jobs can be replaced along with what the consequences are for this radical change. Speaking of repercussions, the overhauling of unskilled labor force is bound to hit blue collar and middle-class families the most, prompting the question of how can this be alleviated.

Human is not only the skills, used for daily routine work but also the mind of creativity and ideas which make any organization to gain competitive edge over other organization. So, with the increase in robotics and automation what will be the role of future leadership, in the attainment of consistent and superior performance by designing such an internal framework of standards and processes which facilitate in engaging and motivating employees to satisfy the customer requirements within business expectations.

In the era of growing phase of robotics, earlier it was used only in the area of aeronautics but now, with the advancement in technology and in artificial intelligence day by day, it has created a need in the management of corporate and business domain. Nowadays, there is a major concern among the business organizations to grow and excel financially, socially, economically and technically. Organizational excellence may be achieved by optimally utilizing the technological resources and consistent improvement within the expert system of the companies.

Robots are infiltrating daily into everyone's lives; there's no ending them. As far as HR personnel are concerned, they need to analyze how this disruption is distressing their respective organizations and need to act accordingly, which involves properly training and reskilling their current employees as well as preparing them for this substantial tectonic shift. people look at these advancements with a mix of distress and anticipation of losing their jobs which will come with new innovative role of human resources. According to a new report from Accenture, in an era when technology is omnipresent, an organization's people will be the competitive differentiators.

Out of the various functions of HR only five functions which includes building a high-performance work system, HR and business strategy, Organizational effectiveness, change management and Employee relations are relatively less vulnerable to automation. Rise of these technology plays a vital role in the industries and service organization and has put a big challenge for the human resource department. In this competitive world, today organizations are committed to improve their technology even at the lower level of HR practitioners to be more cost efficient. Tasks needs to be reinvented and jobs need to be reinvented. Jobs will be taken apart and modified into fundamentally different jobs.

As technology, automation and robotics takes hold, HR experts will need to re-examine their organizations' personnel and the mixture of full-time employees, part-time employees, contractors and machines. Personnel shaping is a novel discipline for human resource with new sets of skills and training to be improvised in attaining the better control towards various challenges and issues with future perspective of adapting changes. The biggest opportunity for HR is to shift toward recognizing that jobs will not be static

The present book has explored and covered the HR Perspective, roles and change in functions towards various challenges and issues with the rise of robotics, automation and artificial intelligence. The primary objective of the book is to create understanding towards changes in functions of HR with the advancement in expert system in addition to the current research by identifying the gaps and bridge them. The book also highlights the role of information technology and futuristic approach towards the leadership. The book promote exchange of information on understanding the use of automation and social media in hiring, innovations in Human Resource Functions with various perspectives, identifying the effect of IT in knowledge sharing and enhancing employee creativity, Technology applications in managing talents and adopting e-HRM systems for the stakeholders connected with the various organizations.

The target audience for this book are composed of Industrial policy makers and researchers from Industry as well as Academia working in the field of organizational behavior, Human Resource & employee relations. The book provides valuable insights and support to practitioners concerned with the management of human resources and organizational development, from the perspectives of their day-to day job activities. This book is the pragmatic source of information for HR managers, administrators, academicians, researchers and students who need to understand the changing HR functions in present scenario with respect to changes in technology and automation.

The book provides various insights with respect to impact of automation on various human resource functions with respect to the various industry and IT sectors. The book reconnoiters the problems and issues of social media recruitment, identifying innovations in human resource, applications of technology in managing talents E- HRM adaptation. The book has chapters related to disruptive technology, social media recruitment, innovation in HRM practices, role of employee empowerment and technology application in managing HR functions. The book has both theoretical and practical implications for industrial policy makers and researchers who wish to embed automation and technology for handling various human resource functions in the organizations. The book comprises of total 9 chapters relating to how HR functions has been transformed with respect to Automation and technology. A chapter-by- chapter brief description is as follows:

Chapter 1 explores the Legitimating disruptive technology – the case of blockchain in the Human Resources sector such research demonstrates that innovator actors seek acceptance for their new technology innovation in different ways, and possibly differently at different stages in the diffusion process. To address these questions, the study draws from current research on legitimation processes and examines the adoption of blockchain in the Human Resources (HR) sector. Blockchain is chosen because it is a radical new technology, often compared with the Internet in the extent to which it is expected to disrupt how organizations operate. Focusing on blockchain allows the study to examine, in real time, the perceptions that organizational actors form about a disruptive technology which is highly unfamiliar. The choice of HR is driven by nature of the sector: in terms of technology adoption, it is more conservative compared with other areas of organizational life. The chapter discuss deliberately on the types of legitimacy considered in the qualifying studies, as well as the kind of strategies that the community deploys to legitimize the technology. The conclusion section will discuss contributions of this research, its limitations identifying thus areas for future research.

Chapter 2 investigates The United Nations sustenance e-government to promote good governance for achieving Sustainable Development Goals – 2030 by fostering environmental, technological, socio-cultural, and political sustainability. Many countries are making sincere efforts to create an ecosystem for fostering e-government to contribute to Sustainable Development Goals. The government is investing a huge amount in e-government projects, but most of them are not able to meet the desired results. It has been reported that most of the e-government is failing due to the non-adoption behavior of employees. This chapter seeks to understand the employee's perspectives by discussing benefits, barriers, and enablers to adopting e-government through a systematic literature review.

Chapter 3 gives insights on Innovations in HRM Practices in Indian Companies the particular chapter examines the as an integral part of every business organization because it provides sustainable competitive advantage to the company. In today's highly dynamic business environment every organization wants to succeed by leveraging their employees' talent. In order to leverage the employees' talent, it is important to follow relevant HRM policies or to continuously introduce innovative HR practices to meet the expectations of the employees. Reviewing the literature and previous research work, this study has tried to find out the various innovative HR practices initiated by Indian organizations till now. The findings will help in guiding how much more innovative practices are still to be initiated in order to attain employee confidence and loyalty for the company. The study is beneficial for business leaders, students, practitioners, and researchers.

Chapter 4 provides the empirical study on The Impact of Social Media on Recruitment in New Age Organizations. The chapter give the insights on the new age world, which has become part of our life, which is a world of rapid innovations and changing technologies. These changes bring new opportunities for organizations to exchange information, news, ideas and work. Attracting and retaining best of employees has become the most crucial tactical problems for the people's department of companies all over the world (Schlechter, Hung & Bussin, 2014; Singh & Finn, 2013). In the current scenario of high competition, the internet has substantially converted the features of recruitment and selection procedure of the businesses. The chapter presents an exploratory research on the impact of social media on recruitment in new age organizations.

Chapter 5 illustrates the Problems and Prospects of Social Media Recruitment. The chapter examines the way talent acquisition has evolved in different forms from attracting personal applications to getting connected with talented candidates through Social networking sites. Recruitment through social networking platforms are putting a significant contribution in analyzing and hiring the right and best talent for an opening and companies can't just ignore the potential and influence of these media platforms. These social platforms connect companies to potential hires and increase visibility by getting them connected to a huge audience. Future of recruitment lies in social media and companies cannot just ignore their presence due to prevailing challenges. It is important to find out viable solutions to the challenge's organizations facing while using social media platforms in talent acquisition. The focus of this chapter is to capture strategies mitigating these challenges and suggest probable and profitable suggestions to companies for better utilization of social networking sites for effective recruitment.

Chapter 6 explores the theoretical framework on Identifying Innovations in Human Resource- Academia and Industry perspective. The chapters climaxes how HR is evolving into more technology based profession because organizations needs to streamline HR processes and reduce administrative burden, reducing administrative cost, compete more effectively with global talent, improve services and access data to the employees and managers, provide real time metrics in order to take on spot decisions for the decision makers and manage the workforce more effectively and enable the HR to transform so it can play more strategic role in the business and operations. The purpose of this chapter is to develop a meaningful debate on the innovations in human resource. Innovations in terms of new ideas, methods and technology to better meet the evolving requirement of the organization and workforce. Anticipating and exploring the future needs and circumstances rather than simply finding some responses to the situation. This chapter highlights challenges and prospects related to innovations in HR.

Chapter 7 explores the effect of information technology and key determinants on knowledge sharing and arriving at employee creativity. The chapter gives insights on Knowledge sharing as an essential parameter for the successful survival of any organization in the modern era. The present chapter aims to investigate the influence of information technology, trust, rewards, leadership, and organizational culture on the knowledge sharing behavior of the employees that ultimately drives employee creativity. To achieve the objectives, primary data was collected using a structured questionnaire at Indore, India with a total of 385 respondents. Structural equation modeling was used to evaluate the relationship between the variables. The findings show that organizational culture followed by leadership was the prominent factor affecting the knowledge sharing behavior of employees. Information technology, trust, and rewards followed next respectively. Employee creativity was found to be significantly affected by knowledge sharing behavior. This chapter not only makes a crucial contribution to the knowledge sharing literature but also benefits the policy-making practices of the organization's top management to facilitate knowledge sharing among the employees.

Chapter 8 this chapter presents a relationship and impact of employee empowerment on organizational trust. The Purpose of this chapter is to analyze the relationship between employee empowerment & organizational trust in IT industry. This study examines the difference between empowerment level and organizational level of male and female employees. The result revealed that the employee empowerment had positive & significant impact on organizational trust. There is no significant difference in the empowerment of IT industry employees. It means that both male and female employees were equally empowered in their jobs. Male employees are more trust from their jobs as compared to female employees.

Chapter 9 provides an introduction and background to Technology applications in managing talents. The chapter explores the how technology applications have changed facets of corporate houses from restructuring organizations to resources to manpower. Artificial Intelligence, Robotic Process Automation and Machine Learning adding an edge to talent acquisition and management. Machine learning tools are primarily being used in acquiring talents and enabling hiring process effectively and efficiently. Today, as every company is moving to a new era of digitalization and data management, managing and mapping talents poses a challenge to C-suite and board level management. The present study highlights the role of technology in managing talents in series of HR process.

I hope the book provides some value information to readers through chapters contributed by various authors. The book will also open the opportunities and avenues for future research in the field of automation and its impact on transforming human resource functions in various industries and corporates.

Chapter 1
Legitimizing Disruptive Technology:
The Case of Blockchain in the Human Resources Sector

Raluca Bunduchi
University of Edinburgh, UK

Aizhan Tursunbayeva
University of Molise, Italy

Claudia Pagliari
University of Edinburgh, UK

ABSTRACT

Previous research demonstrates that actors seek acceptance of innovations during the diffusion and adoption process, but exactly what kind of acceptance they seek and how they go about obtaining this remains unclear. This study interrogated academic databases and Twitter to identify and analyze research and grey literature concerning blockchain adoption in human resources (HR). The data was analyzed through the theoretical lens of pragmatic, moral, and cognitive legitimacy to explore how potential adopters of this innovation conceive its value and alignment with their goals. The results show how pragmatic considerations are emphasized around streamlining current HR processes. Arguments for moral legitimacy are considered far less, while narratives on cognitive legitimacy are absent. The findings also show high levels of conformity around legitimation strategies within the community, with only some evidence of selection and manipulation strategies. The authors discuss the implications of these findings for the adoption and diffusion of blockchain in the HR sector.

DOI: 10.4018/978-1-7998-4180-7.ch001

INTRODUCTION

Human Resources and Blockchain

Emerging technologies are transforming businesses and corporate operations in multiple ways, such as through the use of advanced analytics techniques to analyze and predict consumer behavior or artificial intelligence for high-frequency financial trading. Despite these trends, Human Resources (HR) functions - involving the management and development of employees - appear to be lagging behind other areas of business informatics, although organizations are beginning to recognize their strategic, as well as administrative value (Tursunbayeva, 2018). HR has embraced technologies for routine tasks such as calculating payroll, as well for more strategic activities such as performance management and employee development. The variety and complexity of technologies used in HR departments is also increasing rapidly, partly driven by new vendor offerings and greater budget allocation. In addition to software and information systems, they include platforms, such as cloud infrastructure, as well as techniques such as natural language processing and machine learning. One of the latest innovations to feature in HR is blockchain, a distributed ledger technology which records online transactions and can offer potential benefits for organizations wishing to document and verify assets, processes and supply chains and automate contracts. While the over-hyping of blockchain in recent years has been rightly criticized (Bloomberg, 2017) robust use cases are emerging, which suggest improvements in the efficiency of current business processes, as well as changes in how stakeholders interact with each other. These arise from removing the need for intermediaries, such as banks or estate agents, and by providing immutable data confirming transactions (PWC, 2017). As such, uses of blockchain can now be seen within many areas of the economy, including financial services, healthcare, transport logistics and the public sector. Analysts (e.g. PWC, 2017) have projected that blockchain will have a profound influence on the HR sector, in areas such as talent sourcing, cross-border payments, fraud prevention, cybersecurity and data security, prompting HR innovators and organizational leaders to facilitate its adoption, although robust evidence of these uses and benefits is so far limited.

Legitimation of Technologies

Information systems, management and business researchers have studied in depth the adoption and diffusion of new technology in organizations. Research on innovation diffusion is often split across two camps: the rational choice and the socialized behavior approaches (see also Kaganer et al., 2010). The rational choice approach assumes the decision to adopt technology to be rational and driven by self-interest (see also Fichman, 2004). It draws on theories of the diffusion of innovation (DOI) (Rogers, 2003) and the technology-organization-environment framework (TOE) (DePietro et al., 1990) to examine organizational-level adoption, as well as on technology adoption models (TAM) (Davis, 1989), such as the unified model of technology adoption (UTAUT), to examine individual-level adoption (Venkatesh et al., 2003). The assumption is that individuals and organizations can evaluate the characteristics of the technology, such as its relative advantage (DOI) or its usefulness (TAM) and make a decision on whether to adopt the technology if it is perceived to be advantageous and/or useful. Other factors are also taken into consideration, such as social pressure (UTAUT) or contextual factors, such as the size of the organization in which the technology is to be adopted or the level of competition in the industry (TOE). The biggest factors influencing adoption decisions at both individual and organizational levels are generally the technology and organizational characteristics (Jeyaraj et al., 2006). The rational choice approach assumes that calculations of economic efficiency and performance underpin the adoption of new technologies, such as when assessing the ability of a software product to meet an organizational need. In contrast, the socialized behavior approach, drawing on frameworks such as institutional theory (Mignerat & Rivard, 2009), cultural theory (Leidner & Kayworth, 2006) or structuration theory (Jones & Karsten, 2008), conceptualizes the behavior of organizational actors as being driven by what is socially perceived as legitimate within their environment, rather than by rational calculations (Meyer & Rowan, 1977). Such organizational behavior research finds that the social structures that characterize the macro environment in which organizational actors operate, such as cultural norms (Leidner & Kayworth, 2006), and mimetic, regulative and normative forces (Mignerat & Rivard, 2009), rather than rational calculations of efficiency and effectiveness, shape both the decision to adopt technology and its subsequent use. The socialized behavior approach thus assumes that the need to be perceived as legitimate within their particular environment underpins the decision of organizational actors to adopt new technology.

While most existing studies focus on one or the other of these approaches (see Finchman, 2004; Kaganer et al., 2010), there is also a significant body of research on innovation adoption which seeks to reconcile them, by examining how both rational choice and institutional factors influence adoption decisions (e.g. Sharma et al., 2007; Zhang & Dhaliwal, 2009). Some research suggests that the balance of rational and socialized influences can vary over time; for example at the early stages of diffusion, adoption may be underpinned by rational calculations, while institutional forces reflecting the need for social legitimacy may underpin later stages (Tolbert & Zucker, 1983). Others argue that institutional forces are present from the outset (Swanson & Ramiller, 1997). Recent research suggests that the influence of different institutional forces varies across stages of adoption, with mimetic ones (reflecting cognitive legitimacy) being most influential during the early stages, when uncertainty is high, but regulative and normative ones (reflecting moral legitimacy) becoming more important at later stages, where uncertainty is reduced (Sherer et al., 2016).

STUDY AIM AND RESEARCH QUESTIONS

As already noted, research on the legitimation of information technology has demonstrated that adopters seek acceptance for an innovation in different ways, which can vary across stages in the diffusion process. However, *the strategies through which they do so are not yet understood*, nor *whether these strategies vary across stages*. This study set out to address this research gap by examining the adoption of blockchain in the HR sector. Blockchain is chosen for this study because it is a radical new technology (Iansiti & Lakhani, 2017), which has been compared to the Internet in the extent to which it is expected to disrupt how organizations operate (Tapscot & Tapscot, 2016). Outside the financial services sector, where the first blockchain application emerged, the technology is still in its early stages of diffusion (Du et al., 2019). Recent reviews have identified very few commercial-grade blockchain applications (Hughes *et al.*, 2019) or documented cases of large-scale implementations, meaning that there is limited understanding of its overall value (Pan et al., 2020). With limited evidence to support the development of a clear business case for adoption, most blockchain adoptions are currently driven by the perceptions actors have about the technology and its benefits. Focusing on the blockchain thus allows the study to examine, in real-time, the perceptions that organizational actors form about a disruptive technology that is highly unfamiliar. The choice of HR is driven by the nature of the sector: in terms of technology adoption, it is more conservative compared with other areas of organizational life. By focusing on the adoption of a radical technology in a relatively conservative area, we explore a sector where the need of innovator actors to legitimize a disruptive new technology to encourage its adoption is more intense (in comparison with other sectors).

THEORETICAL FRAMEWORK

Legitimacy - Types and Strategies

Achieving organizational legitimacy involves aligning to what is seen as *"desirable, proper, or appropriate within some socially constructed system of norms, values, beliefs, and definitions"* (Suchman, 1995; p. 574). Different forms of organizational legitimacy are identified in existing research, applied in particular to understand organizational behavior in general (Oliver, 1991), the diffusion of new practices (Greenwood et al., 2002), or entrepreneurial behavior in particular (Zimmerman & Zeitz, 2002). The most commonly used legitimacy framework applied across those fields is Suchman's (1995), which differentiates between organizational legitimacy depending on their underlying behavioral dynamics. Three broad types of legitimacy are identified: pragmatic, moral, and cognitive. *Pragmatic legitimacy* assumes an organization to be desirable based on its utility to the audience and is based on *"the self-interest calculations"* of an organization's stakeholders (Suchman, 1995, p. 578). Such self-interest calculations involve the assessment of the organization's expected value; its degree of alignment with the stakeholders' goals and interests; or its good character, i.e. whether the organization is honest, trustworthy, decent and wise. *Moral legitimacy* assesses the desirability of an organization based on its alignment with the normative beliefs embedded within the audience's institutionalized value system and is based on evaluating organizational actions as being *"the right thing to do"* (Suchman, 1995, p. 579). Such evaluations may concern the organization's consequences and outputs; its techniques and procedures; its structural characteristics and the categories it belongs to; or its leaders and representatives. *Cognitive legitimacy* assesses the desirability of an organization based on its alignment with the cultural model embedded within the audience's institutionalized value system. Such cognitive assessments of whether the organization is valid may be based either on its inevitability as the organization is *"taken for granted"* within a particular context so that alternatives are unthinkable; or its comprehensibility as the organization aligns with the prevalent cultural model thus being comprehensible to its audience (Suchman, 1995).

Suchman (1995) differentiates between different actions involved in the legitimization process of an organization: acquiring, repairing and maintaining. The triggers for legitimacy-seeking actions are thus assumed to be when an organization is new and lacks legitimacy (for acquiring actions), when an organization goes through a crisis that has damaged its legitimacy and it is forced to repair it (for repairing actions), or when the organization is simply seeking to maintain the status quo (for maintaining actions). Seeing our focus on new technology, we focus here on acquiring actions. Each action is associated with different kinds of legitimization strategies. Suchman (1995) differentiates three such legitimization strategies: *conforming* to the expectations of key stakeholders, *selecting* a favorable group of stakeholders with aligned expectations and mindset to target, and *manipulating* the context in which stakeholders are embedded to alter their mindsets and behaviors. Zimmerman and Zeitz (2002) also identify *creation* strategies where innovator actors become a pioneer who actively designs new explanations of social reality to launch a new supportive environment. A plethora of studies both in the entrepreneurship and organizational fields have examined the ways in which innovator actors deploy such legitimization seeking strategies to achieve different types of legitimacy at different times (e.g. Fisher et al., 2017; Deephouse et al., 2016).

Legitimization Seeking and Technology Adoption

Within information systems (IS), legitimization has been conceptualized as a critical mechanism to ensure the diffusion of a new technology within a community (Kaganer et al., 2010). For example, Swanson and Ramiller (1997) argue that for a new technology, to diffuse within a new community, the community needs to form a coherent interpretation about this new technology – what they call organizing vision – which plays in a critical role in legitimizing of the new technology and thus mobilizing resources to generate interest and support its adoption (see also Currie, 2004). Empirical studies find that lack of congruence between the visions developed by different groups of actors such as vendors, industry analysts and academic institutions in software sector (Currie, 2004), and policy makers, technology vendors, researchers and academic institutions, clinicians and management consultants in healthcare industry (Greenhalgh et al., 2012) hamper diffusion of new technologies in these sectors. Existing organizing vision IS research tends to emphasize the need for coherence between the visions developed by the different members of the community involved in the development and use of the new technology to ensure the new technology is seen as legitimate and is thus adopted (Wang & Swanson, 2007). In contrast, less is explored about the mechanisms through which such legitimization is sought or the types of legitimacy that actors seek to ensure the acceptance of the new technology. In an extensive study of the diffusion of a new technology within healthcare, Kaganer et al. (2010) documented a plethora of actions – whose pattern forms specific types of legitimization strategies - that technology vendors deployed to legitimize their new technology within the community and ensure its adoption. Their study demonstrated the importance of legitimacy seeking behavior by technology vendors to ensure the adoption of a new technology. This research aims to further explore legitimacy seeking behavior for new technology by identifying the types of legitimization strategies and legitimacies that innovative actors deploy and seek respectively to ensure the adoption of a new technology.

In conclusion, while there is a wide body of evidence that legitimacy matters in the adoption and diffusion of new technologies, and that technology vendors engage in a wide range of legitimacy seeking behavior, it is unclear which kind of strategies are deployed, and what kinds of legitimacies are sought by the range of actors involved. To explore these issues, we focus here on the early stages of the diffusion of a radical technology – blockchain - within a particular community – HR sector.

METHODS

To study the process involved in the legitimization of blockchain within the HR sector, we conducted a scoping review of (1) academic literature, alongside a review of (2) grey literature posted on Twitter. This type of scoping review is commonly used to examine emerging topics that are poorly understood, where research is at an early stage, or where pertinent knowledge is being generated outside academia (e.g. Tursunbayeva et al., 2018). The analysis of academic and grey literature will allow us to track the perceptions of range of stakeholders involved - often potential adopters, but also research experts - over time.

- **Scoping Academic Literature:** A search query was created from blockchain and HR related keywords (i.e. (("Blockchain" OR "Block chain") AND "Human Resource*")). The query was used to interrogate three interdisciplinary online literature databases including Web of Science Core Collection (WoS), ScienceDirect and Scopus. The references of all articles that were included for the full analysis were also snowballed. In addition, the articles citing the qualifying articles were checked to identify additional potentially relevant publications.
- **Scoping Grey Literature:** Using Blockchain and HR related hashtags (i.e. "#HR #Blockchain" and "#Humanresources #Blockchain") we searched Twitter (using "Twitter advanced search" function). The search period laid between 21/03/2006, the date when Twitter was created, and 9/12/2018. The same tweet published by two or more authors was not considered a duplicate and was taken into consideration. The full-texts of publications identified via the Twitter hashtag searches were located and analyzed.

The findings from both academic and grey literature were extracted into a predefined excel spreadsheet. They were then open-coded (e.g. on HR management practices) or grouped into the predefined categories (e.g. pragmatic, moral or cognitive legitimacy).

FINDINGS

Overview of Academic and Grey Literature on HR and Blockchain

Searching selected literature databases brought 33 publications. Five publications remained after removing duplicates and non-relevant returns. One manually found article was added to these (see Figure 1). Four publications were published in 2017, while two in 2018. Five publications are IS related conference papers, and one is a journal article. The articles were published by 24 independent authors from China (n=11), Russia (n=6), Ireland (n=4), Italy (n=2), and the USA (n=1). None of these authors tweeted about their research on Twitter. The study designs were equally distributed between technical papers, discussion papers, and case studies. None of the studies drew on any theoretical or methodological frameworks. HR practices discussed in the studies related to the recording of verifiable contributions (n=2), employee selection, handling temporary employee contracts, HR information management, and employee benefits (all n=1).

220 tweets containing the hashtags of interest were extracted from a Twitter search. Specifically, we found 159 tweets containing "#HR #blockchain" and 61 tweets containing "#humanresources #blockchain". After removing duplicates (i.e. tweets by the same author), non-relevant to both HR and Blockchain tweets, tweets not in English or Italian, live tweets, videos and podcasts 52 tweets containing news and/or grey literature articles remained for full analysis (see Figure 1). Three of these tweets were posted in 2016, 12 in 2017, and 37 in 2018. These tweets were tweeted from 48 individual Twitter user accounts.

Figure 1. Approach to search and analysis of academic and grey literature

Literature	Identified	Included for final analysis
Academic ("Blockchain" OR "Block chain") AND "Human Resource*")	WoS/ScienceDirect/Scopus=33 Manually added=1	6
Grey ("#HR #Blockchain" and "#Humanresources #Blockchain")	Tweets=220	52

Analysis of grey literature revealed expectations that Blockchain will disrupt *Recruitment* HR practice (n=43). Most tweets were related to news items or articles on how blockchain will transform recruitment, mostly by allowing candidates to have validated records on their previous employment or educational certifications and enabling potential employers to access them thus eliminating the need to verify these with third parties. The next most frequently mentioned HR practice was *Payroll* (n=10) including also *benefits management* practice. Here it was mentioned that employees and especially freelance employees with the help of blockchain can be paid immediately with cryptocurrencies, facilitating cross-border financial transactions and eliminating long waiting times associated with traditional money transfers. Grey literature also mentioned that blockchain can affect employee Performance management, Training and Development including also new employees' onboarding, as well as increasing overall compliance of organizations associated with personnel data management.

Looking at the stakeholders involved in the blockchain HR community, the results show that the targeted stakeholders seen to benefit the most from blockchain are *organizations* as employers, HR managers or recruiters (n=36), while only 18 studies mentioned *candidates* or current employees.

Excluding the hashtags used for the search, other hashtags used in the tweets identified were #fintech (n=31), #AI (n=27), #IoT (n=23), #futureofwork (n=22), #hrtech (n=19), #recruitment (n=16) #bitcoin (n=15), #tech (n=15), #bigdata (n=14), #innovation (n=13), #finance (n=12).

While some studies from both academic and grey literature focused only on some specific types of legitimacy, other studies discussed several of them together.

THE LEGITIMACY OF BLOCKCHAIN IN THE ACADEMIC LITERATURE

Pragmatic Legitimacy

Academic literature mostly considered the pragmatic legitimacy of blockchain and its applications in HR, with studies mainly emphasizing its value for organizations and, to a lesser extent, its alignment with their goals. For organizations, the main value from using blockchain in HR concerned the enhancement of the security, reliability, and verifiability of information about applicants, which can be beneficial both to individuals (as candidates) and to organizations (as employers). Thus, for example, candidates who have their diplomas, certificates or other documents on the blockchain system can have these automatically verified by the relevant organizations that issued them and passed to potential employers. Existing research emphasized that stakeholders reap such value mostly during employee selection and recruitment.

One study (O'Leary et al., 2017) also argued that blockchain creates value for organizations by reducing the social loafing phenomenon, when individuals take advantage of group work as their individual contributions to group outputs are difficult to identify. Social loafing has negative connotations as it is seen to reduce individual motivations (George, 1992). Blockchain facilitates the establishment of the transparent, visible, and verifiable approach for capturing employees' work. Transparency ensures that individual contributions are recognized, acknowledged, and rewarded, thus reducing the incidence of social loafing and improving individual motivation to actively participate in the group work.

Additionally, blockchain was seen as helping organizations and employees to achieve their privacy goals by protecting sensitive information. For example, Ying et al. (2018) described how an organization launched an innovative employee benefits program that, with the help of blockchain, enabled employees to be identified by suppliers without revealing their personal information. Blockchain was also seen as enabling organizations to avoid costly and inefficient institutional intermediaries by shifting the trust established by this intermediary institutional authority to the blockchain network.

For pragmatic legitimacy the key strategy to acquire legitimacy was implicitly conforming, with authors emphasizing the inherent qualities of the technology which provide expected benefits and align with existing goals, thus demonstrating that the technology conforms with current expectations. There was no evidence of selection strategies – for example in the form of explaining that the technology benefits some organizations or candidates over others, nor manipulation or creation strategies, in that the adopters need to change or alter their expectations, or goals to extract the most value of the technology.

Moral Legitimacy

Some academic publications (e.g. Pinna & Ibba, 2017) also considered the adoption of blockchain in HR from a normative standpoint. For example, using blockchain when allocating temporary employment contracts, was seen as a means of protecting both employers' and employees' rights, thus aligning with current normative expectations (the technology can be used by competent authorities to monitor the execution of temporary employment conditions for verifying for example that their employment conditions are correctly fulfilled and any taxation related illegal actions are absent). The focus here is on the alignment between the outputs of using the technology and current normative expectations.

Similarly, with pragmatic legitimacy, the focus seems to be on conforming rather than manipulation strategies, with the emphasis on alignment with current expectations of organizations to conform to regulative norms. Implicitly however, the studies consider selection strategies through emphasizing the ability of organizations to monitor temporary contracts, hence suggesting that blockchain may be likely to be more accepted by organizations where temporary contracts are significant to ensure their compliance with normative rules.

THE LEGITIMACY OF BLOCKCHAIN IN GREY LITERATURE

Pragmatic Legitimacy

In line with the academic literature, the grey literature mostly considered the pragmatic legitimacy of blockchain and its applications in HR. Thus, it proposes that blockchain can create significant value for organizations by streamlining their current employee selection process, and for candidates by streamlining the work search and application processes. The argument here is that blockchain's ability to ensure that data entered once is unamendable, which makes the storage, validation, and verification of employees' data, such as relevant work experience and qualifications, cost-effective, fast, efficient and accurate for organizations. Blockchain also promises to match the right job with the right candidates, thus making the expensive fees of traditional job portals and recruiters obsolete and allowing the companies to find candidates directly. So-called "smart contracts" enabled by blockchain are forecasted to be executed automatically when conditions are filled, meaning that payments can be made without the involvement of any intermediary banks or third parties, without any delays, and in the form of cryptocurrencies. They also promise to streamline a lot of the onboarding process, especially for high volume, high turnover positions. For global employees, blockchain is seen as a means to streamline the process involved in seeking to adhere with local tax jurisdictions, and to manage cross-border expense procedures by including the ability to code rules and limitations into the blockchain, potentially eliminating the need for a signature or authorization process. Moreover, for a globalized workforce, the ability to prove identity, authenticate right to work and to even remain in a country will be critical for both employees and employers in the future.

For job candidates, blockchain is depicted as a living, organic Curriculum Vitae that provides indisputable evidence of an individual career in real time. Candidates should also be able to create an easily accessible record of every job application, interview process, and outcome. This alone should enable candidates to gradually revolt against having to constantly duplicate applications for the same role with dozens of individual companies, as in essence, every application usually requires the exact same information and documentation and lengthy repetitive processes. Moreover, some authors also mentioned that Blockchain can give more ownership of data to employees or individuals.

Such arguments center around legitimizing the technology by demonstrating the alignment between the technology and the current processes and practices – and expectations of value – for both organizations and candidates. In common with the academic research, the focus here is on conforming strategy to demonstrate pragmatic legitimacy.

Contrary to academic research, grey research emphasizes also the disruptive aspect of blockchain to current practices. Blockchain is, for example, envisioned to revolutionize the short-term recruitment sector by creating decentralized mechanisms for employers to access and rank prospective employees. Moreover, for contract workers data about payroll, taxes, expenses can be immediately available via the blockchain. Such emphasis on the revolutionary aspect of technology suggests efforts to encourage organizations to consider the value of novel ways of organizing their HR practices, thus being indicative of manipulation strategies to alter current mindsets of how HR should be managed within organizations, and thus the value that blockchain can bring within these potentially altered settings.

Moral Legitimacy

In common with academic research, the grey literature emphasizes the beneficial outcomes that using blockchain can contribute in HR, such as by enabling organizations and job candidates to protect privacy and documenting transactions, contracts and agreements to establish a permanent source of truth for organizations and candidates. Blockchain is associated with more transparent, decentralized processes, thus making it easier to spot fraud in employee credentials. Searching for records and validating data in order to audit and prove compliance is also seen as faster and more accurate, leading to a more private and less bias-prone recruitment process which aligns with normative expectations for fairness and privacy in HR practices. However, some authors also questioned the confidentiality of candidates' job search process via blockchain recruitment platforms because by exposing this search to their existing employers, employees can potentially jeopardize their current jobs.

Cognitive Legitimacy

Only one source identified in the grey literature considers aspects related to the cognitive legitimacy (specifically to the lack of cognitive legitimacy), although it does not mention any actions to gain such legitimacy. Blockchain is seen a foundational technology that has the potential to create a new basis for our economic and social systems. Such foundational technologies however require a large amount of time to become embedded into economic and social infrastructure. Thus, it is estimated that the revolutionary impact of blockchain on HR practice will take at least 5-10 years from now because of the magnitude of change that blockchain-enabled processes require, compared with the current "taken for granted" HR practices.

DISCUSSIONS, CONCLUSION AND FUTURE RESEARCH DIRECTIONS

This exploratory study draws on published academic and grey literature to document how adopters seek to legitimize the usage of blockchain technology in the HR sector. Our analysis shows that academic research on HR and blockchain is scarce, albeit growing, and mostly atheoretical. Most scholarly research on this topic has been undertaken from an information systems perspective. While this elucidates technical aspects of blockchain, and how the technology could be potentially developed, implemented or used by HR functions, it leaves open questions around whether and how blockchain can help HR functions to operate or manage employees more effectively.

Our explorative analysis revealed a focus on pragmatic legitimacy within the existing academic and grey literature. Of the academic papers reported, most were aimed at demonstrating the value of blockchain to potential adopters and aligning these with organizational goals around streamlining current HR practices and processes such as storage, validation and verification of data during recruitment process. There was relatively little consideration of how blockchain offers new or better ways of organizing HR practices, such as through offering transparent and verifiable approaches for capturing productivity. Such a strong focus on pragmatic legitimacy seems to be aligned with the body of research suggesting that in the early stages of an innovation, adoption is driven by rational calculations, with institutional norms and socialized behavior playing a stronger role during the later stages (Tolbert & Zucker, 1983). However, further research is needed to verify this.

For job candidates, blockchain is depicted as a living, organic Curriculum Vitae that provides indisputable evidence of an individual career in real time. Candidates should also be able to create an easily accessible record of every job application, interview process, and outcome. This alone should gradually enable candidates to avoid having to constantly duplicate applications for the same role with dozens of individual companies, as in essence, every application usually requires the exact same information and documentation and lengthy repetitive processes. Moreover, some authors also mentioned that Blockchain can give more ownership to employees or individuals for their personal data (e.g. O'Leary et al., 2017).

Such arguments center around legitimatizing the technology by demonstrating the alignment between the technology and the current processes and practices – and expectations of value – for both organizations and candidates. Similar with the academic research, the focus here is on conforming strategy to demonstrate pragmatic legitimacy.

Contrary to academic research, the grey literature emphasizes also the disruptive aspect of blockchain to current practices. Blockchain is, for example, envisioned to revolutionize the short-term recruitment sector by creating decentralized mechanisms for employers to access and rank prospective employees. Moreover, for contract workers, data about payroll, taxes and expenses can be immediately available via the blockchain. Such emphasis on the revolutionary aspect of technology suggests efforts to encourage organizations to consider the value of novel ways of organizing their HR practices, thus being indicative of manipulation strategies to alter current mindsets of how HR should be managed within organizations, and thus the value that blockchain can bring within these potentially altered settings.

There was much less emphasis on seeking moral legitimacy, although this was more obvious in the grey literature, than in academic articles. Moral considerations mostly concerned the expectation that blockchain would help to support privacy and fair recruitment processes. This suggests that socialized behavior does play a role in the early stages of this innovation, even not if to the extent seen in other research (e.g. Swanson & Ramiller, 1997). The observation of moral, rather than cognitive legitimization narratives also contrasts with existing research focused on the early stages of technology adoption (Sherer et al., 2016).

Indeed, amongst the eligible studies there was no obvious discussion of cognitive legitimacy. While this observation is based on only six scholarly studies it is important also to note the context of study – a disruptive innovation in what a relatively conservative sector. It may well be in such situations that to overcome the resistance to adoption, the focus is on expected value and alignment with existing goals, rather than cognitive alignment, compared with other sector sectors where cognitive norms favor engaging with radical technology.

The findings also revealed the strategies different communities use to legitimize this technology: with conformity organizational strategy primarily seen in the academic research, and the grey literature also demonstrating selection and manipulation. This is an interesting finding, as most current research on technology adoption suggests conformance as the key strategy for legitimizing an innovation (see Mignerat & Rivard, 2009), despite effort by entrepreneurship and organizational researchers to demonstrate a wider range of legitimation strategies. The lack of creation strategies may also be associated with the context of investigation – HR is still often seen as an administrative function, as a cost center, with organizations preferring to invest in areas that are seen as more critical (Tursunbayeva, 2018) such as supply chain, customer relations or finance. Thus, HR is unlikely to be at the forefront of adopting radical practices that require creativity.

Findings mostly arising from the grey literature, indicate that organizational leaders and HR professionals are the main target audience for legitimation efforts, while also acknowledging employees/candidates as stakeholders. By and large, existing IS adoption research mostly ignores the variety of potential adopters, and rarely consider both individuals and organizations within the same study. We therefore plan to undertake a follow-up study involving expert interviews, to (1) clarify the legitimation strategies that emerged from the academic and grey literature we analyzed, and (2) explore the conditions under which different strategies are deployed to demonstrate to organizations and/or employees/candidates that the technology is useful, aligned with their norms and/or culturally valid. The follow-up study also aims to disentangle the types of strategies deployed to seek legitimacy between organizations and candidates, as well as explore the interactions between them.

Finally, none of the literature we analyzed considered or described any negative outcome or unintended consequence of HR in blockchain, although some considered practical challenges such as how to enable candidates to search for new jobs without disclosing this to their current employers (Balashova, 2018). This implies the need for future research to investigate the unanticipated obstacles and unintended consequences of these innovations.

Key messages for HR practitioners are that, despite the widespread discourse around blockchain as a disruptive innovation, strategies for encouraging its adoption and diffusion tend to focus on conformance with existing expectations and economic benefits for users. This suggests that organizations wishing to facilitate adoption should place less emphasis on its role as a disruptor and more on its ability to align with and strengthen existing practices. When drawing conclusions about innovations, it is important to be mindful of their maturity. As blockchain technology diffuses into HR, it is likely that pragmatic rationales will be supersede by moral ones, or conforming strategies give way to creative or manipulative ones, as expectations change and prior assumptions are disproven.

In conclusion, we hope that this study will stimulate more conceptual, theoretical and empirical work on how the use of blockchain is being legitimized in HR, as well as contributing to academic research on the legitimation of information technologies.

REFERENCES

Balashova, A. (2018). Unblocking blockchain recruitment. *Future Times*.

Bloomberg, J. (2017). Eight reasons to be skeptical about blockchain. *Forbes*.

Currie, W. L. (2004). The organizing vision of application service provision: A process-oriented analysis. *Information and Organization*, *14*(4), 237–267. doi:10.1016/j.infoandorg.2004.07.001

Davis, F. D. (1989). Perceived usefulness, perceived ease of use, and user acceptance of information technology. *Management Information Systems Quarterly*, *13*(3), 319–340. doi:10.2307/249008

Deephouse, D. L., Bundy, J., Tost, L. P., & Suchman, M. C. (2016). Organizational legitimacy: Six key questions. In R. Greenwood, C. Oliver, T. Lawrence, & R. Meyer (Eds.), *The SAGE handbook of organizational institutionalism* (pp. 27–54). Sage.

DePietro, R., Wiarda, E., & Fleischer, M. (1990). The context for change: organization, technology and environment. In L. G. Tornatzky & M. Fleischer (Eds.), *The Process of Technological Innovation* (pp. 151–175). Lexington Books.

Du, W., Pan, S. L., Leidner, D. E., & Ying, W. (2019). Affordances, experimentation and actualization of FinTech: Blockchain implementation study. *The Journal of Strategic Information Systems*, *28*(1), 50–65. doi:10.1016/j.jsis.2018.10.002

Fichman, R. G. (2004). Going beyond the dominant paradigm for Information Technology innovation research: Emerging concepts and methods. *Journal of the Association for Information Systems*, *5*(8), 314–355. doi:10.17705/1jais.00054

Fisher, G., Kotha, S., & Lahiri, A. (2016). Changing with the Times: An Integrated View of Identity, Legitimacy, and New Venture Life Cycles. *Academy of Management Review*, *41*(3), 383–409. doi:10.5465/amr.2013.0496

Fisher, G., Kuratko, D. F., Bloodgood, J. M., & Hornsby, J. S. (2017). Legitimate to whom? The challenge of audience diversity and new venture legitimacy. *Journal of Business Venturing*, *1*(32), 52–71. doi:10.1016/j.jbusvent.2016.10.005

George, J. M. (1992). Extrinsic and Intrinsic Origins of Perceived Social Loafing in Organizations. *Academy of Management Journal*, 191–202.

Greenhalgh, T., Procter, R., Wherton, J., Sugarhood, P., & Shaw, S. (2012). The organising vision for telehealth and telecare: Discourse analysis. *BMJ Open*, *2*(4), 1–12. doi:10.1136/bmjopen-2012-001574 PMID:22815469

Greenwood, R., Suddaby, R., & Hinings, C. R. (2002). Theorizing change: The role of professional associations in the transformation of institutionalized fields. *Academy of Management Journal*, *45*(1), 58–80.

Hughes, L., Dwivedi, Y. K., Misra, S. K., Rana, N. P., Raghavan, V., & Akella, V. (2019). Blockchan research, practice and policy: Applications, benefits, limitations, emerging research themes and research agenda. *International Journal of Information Management*, *49*, 114–129. doi:10.1016/j.ijinfomgt.2019.02.005

Iansiti, M., & Lakhani, K. R. (2017). The truth about blockchain. *Harvard Business Review*, *95*(1), 118–127.

Jeyaraj, A., Rottman, J. W., & Lacity, M. (2006). A review of the predictors, linkages, and biases in IT innovation adoption research. *Journal of Information Technology*, *21*(1), 1–23. doi:10.1057/palgrave.jit.2000056

Jones, M. R., & Karsten, H. (2008). Giddens's structuration theory and information systems research. *Management Information Systems Quarterly*, *32*(1), 127–157. doi:10.2307/25148831

Kaganer, E., Pawlowski, S. D., & Wiley-Patton, S. (2010). Building legitimacy for IT innovations: The case of computerized physician order entry system. *Journal of the Association for Information Systems*, *11*(1), 1–33. doi:10.17705/1jais.00219

Leidner, D. L., & Kayworth, T. (2006). A review of culture in Information Systems research: Towards a theory of information technology culture conflict. *Management Information Systems Quarterly*, *30*(2), 357–399. doi:10.2307/25148735

Meyer, J. W., & Rowan, B. (1977). Institutionalized Organizations: Formal Structure as Myth and Ceremony. *American Journal of Sociology*, *83*(2), 340–363. doi:10.1086/226550

Mignerat, M., & Rivard, S. (2009). Positioning the institutional perspective in information systems research. *Journal of Information Technology*, *24*(4), 369–391. doi:10.1057/jit.2009.13

O'Leary, K., O'Reilly, P., Feller, J., Gleasure, R., Li, S., & Cristoforo, J. (2017). Exploring the Application of Blockchain Technology to Combat the Effects of Social Loafing in Cross Functional Group Projects. *OpenSym '17: Proceedings of the 13th International Symposium on Open Collaboration*, 1–8. 10.1145/3125433.3125464

Oliver, C. (1991). Strategic responses to institutional processes. *Academy of Management Review*, *16*(1), 145–179. doi:10.5465/amr.1991.4279002

Pan, X., Pan, X., Song, M., Ai, B., & Ming, Y. (2020). Blockchain technolgoy and enterprice operational capabilities: An empirical test. *International Journal of Information Management*, *52*.

Pinna, A., & Ibba, S. (2017). A blockchain-based Decentralized System for proper handling of temporary Employment contracts. *Intelligent Computing: Proceedings of the 2018 Computing Conference*, *2*.

PWC. (2017). *How will blockchain technology impact HR and the world of work?* Author.

Rogers, E. M. (2003). *Diffusion of Innovations*. Free Press.

Sharma, A., Citurs, A., & Konsynski, B. (2007). Strategic and institutional perspectives in the adoption and early integration of Radio Frequency Identification (RFID). *Proceedings of 40th Annual Hawaii International Conference on System Sciences (HICSS'07)*. 10.1109/HICSS.2007.502

Sherer, A. A., Meyerhoefer, C. D., & Peng, L. (2016). Applying institutional theory to the adoption of electronic health records in the U.S. *Information & Management, 53*(5), 570–580. doi:10.1016/j.im.2016.01.002

Suchman, M. (1995). Managing legitimacy: Strategic and Institutional Approaches. *Academy of Management Review, 20*(3), 571–610. doi:10.5465/amr.1995.9508080331

Swanson, E. B., & Ramiller, N. C. (1997). The organizing vision in information systems innovation. *Organization Science, 8*(5), 458–474. doi:10.1287/orsc.8.5.458

Tapscott, D., & Tapscott, A. (2016). *Blockchain revolution: how the technology behind bitcoin is changing money, business, and the world*. Penguin Random House.

Tolbert, P. S., & Zucker, L. G. (1983). Institutional sources of change in the formal structure of organizations: The diffusion of civil service reform, 1880-1935. *Administrative Science Quarterly, 28*(1), 22039. doi:10.2307/2392383

Tursunbayeva, A. (2018). *Human Resource Management Information Systems in Healthcare. Processes of development, implementation and benefits realization in complex organizations*. Franco Angeli.

Tursunbayeva, A., Di Lauro, S., & Pagliari, C. (2018). People analytics—A scoping review of conceptual boundaries and value propositions. *International Journal of Information Management, 1*(43), 224–247. doi:10.1016/j.ijinfomgt.2018.08.002

Venkatesh, V., Morris, M. G., Davis, G. B., & Davis, F. D. (2003). User acceptance of information technology: Toward a unified view. *Management Information Systems Quarterly, 27*(3), 425–478. doi:10.2307/30036540

Wang, P., & Swanson, E. B. (2007). Launching professional services automation: Institutional entrepreneurship for information technology innovations. *Information and Organization, 17*(2), 59–88.

Ying, W., Jia, S., & Du, W. (2018). Digital enablement of blockchain: Evidence from HNA group. *International Journal of Information Management, 39*, 1–4. doi:10.1016/j.ijinfomgt.2017.10.004

Zhang, C., & Dhaliwal, J. (2009). An investigation of resource-based and institutional theoretic factors in technology adoption for operations and supply chain management. *International Journal of Production Economics, 120*(1), 252–269. doi:10.1016/j.ijpe.2008.07.023

Zimmerman, M. A., & Zeitz, G. J. (2002). Beyond survival: Achieving new venture growth by building legitimacy. *Academy of Management Review, 27*(3), 414–431. doi:10.5465/amr.2002.7389921

Chapter 2
E–Government Adoption Among Employees:
A Systematic Review–Derived Conceptual Framework

Preeti Bhaskar
ICFAI University, Dehradun, India & University of Technology and Applied Sciences, Ibra, Oman

Muddu Vinay
ICFAI University, Dehradun, India

Amit Joshi
ICFAI University, Dehradun, India

ABSTRACT

The United Nations have sustenance e-government to promote good governance for achieving Sustainable Development Goals – 2030 by fostering environmental, technological, socio-cultural, and political sustainability. Many countries are making sincere efforts to create an ecosystem for fostering e-government to contribute to Sustainable Development Goals. The government is investing a huge amount in e-government projects, but most of them are not able to meet the desired results. It has been reported that most of the e-government is failing due to the non-adoption behavior of employees. This chapter seeks to understand the employee's perspectives by discussing benefits, barriers, and enablers to adopting e-government through a systematic literature review.

DOI: 10.4018/978-1-7998-4180-7.ch002

INTRODUCTION

The United Nations (UN) have sustenance e-government to promote good governance for achieving sustainable Development Goals – 2030 by fostering environmental, technological, socio-cultural, and political sustainability (Ziemba, 2018; United Nations, 2018, 2015). UN makes use of the e-government development index framework (EGDI) to give performance ratings to member states of United Nations on three components i.e. availability of online services, telecommunications infrastructure and human resources (United Nations Department of Economic and Social Affairs, 2003). Many countries are making sincere efforts to create an ecosystem to foster e-government in the previously mentioned three dimensions. Several researchers around the world have discussed the growth of e-government at the local, state, global, and international levels (Wang and Feeney, 2016; West, 2000; Franke and Eckhardt, 2014; Ingrams et al., 2018). E-government's goals include sharing information and providing quality services at minimal cost. It enhances communications with citizens, businesses, government employees and makes the system transparent and efficient. Numerous investigators have reported the advantages of e-government like increased transparency and combat corruption (Ismail et al., 2020); costs reduction (Suzuki and Suzuki, 2020); time-saving (Gilbert et al., 2004); economic growth (Azim et al., 2020); and sustainable development (Othman et al., 2020). The benefits cannot be reaped to full extent until employees adopt e-government. Employees are frontline workers in delivering e-government services to the citizen in an efficient and effective manner. Most of the e-government is not able to obtain desired results because of the non-adoption behavior of employees (Stefanovic et al., 2016; Rowley, 2011). Employees face multiple barriers in delivering services to all stakeholders. Stakeholders are optimistic about e-government but they are more apprehensive about various barriers. Quite a few researchers have identified the e-government barriers like technology barriers (Al-Refaie & Ramadna, 2020), organizational barriers and financial barriers (Zhang et al., 2005), lack of resources and lack of willingness (Rehouma, 2020), organizational structure and cultural (Batara et al., 2017) and security and anxiety (Rana et al., 2013). These barriers restrict the success of e-government and need to be supported by enablers which will push its successful implementation. Literature shows various enablers that stimulate employees to adopt e-government include heterogeneous network interconnections and security (Kettani, 2014) infrastructure and legal framework (Srivastava & Teo, 2004), computer efficacy (Rana et al., 2013) and employees socio-demographic characteristics (Dukić ey al., 2017).

OBJECTIVES AND METHODOLOGY

This paper seeks to understand the employee's perspectives by discussing benefits, barriers, and enablers in adopting e-government through a systematic literature review (SLR). The author(s) have consolidated the existing review of literature on "e-government adoption among employees". This research has systematically performed the bibliometric analysis of the studies to explore the barriers faced by employees in adopting e-government, to identify enablers of e-government adoption among employees and to enumerate benefits of e-government adoption among employees. The research also proposes systematic review-derived conceptual framework for adopting e-government among employees. Findings and results of the study are extremely important for policy and decision makers to formulate the strategic framework to adopt e-government by the employees for the efficacious operation of e-government. The authors have used the steps involved in Preferred Reporting Items for Systematic Reviews and Meta-Analyses (PRISMA) (Figure – 1) by clearly mentioning the data search, inclusion criteria and exclusion criteria for the study under consideration. Many researchers have used the PRISMA approach for conducting SLR (Mustafa et al., 2020; Mahmood et al., 2014; Alzahrani et al., 2017; Batubara et al., 2018)

In step 1, the authors conducted the review in SCOPUS database because it has a wide coverage of peer-reviewed high-quality research papers. Published studies were found by searching the SCOPUS database through the university library system to meet the research objectives. The initial search was performed using different keyword combinations. "e-government adoption among employees", "factors influencing employee adoption of e-government", "adoption of e-government among employees". Furthermore, there were also no restrictions put on the language and publication year for retrieving the documents. The search resulted in 78 documents that consist of 35 articles, 27 conference papers, 3 review paper, 1 book, 7 book chapters, and 5 conference review (Table- 1).

In step 2, a visualizing examination was conducted for the conditions of inclusion for the articles for further reviews. Unrelated paper, duplicated paper, book, book chapters, conference paper and editorial were eliminated. We imposed few conditions for the inclusion of articles for further study. The first condition is that the article should be published in English language; second condition is it should not be the part of "article in press", "conference review", "conference paper", "book chapter" and "book". Third condition is that the article should discuss and examines e-government adoption among employees. The papers were omitted if they did not discuss any of the three conditions. Based on the above-specified inclusion criteria all papers were screened carefully. Finally, 27 (n = 78) articles met the three conditions that were selected for further study (**Figure. 1**).

Figure 1. Systematic literature review

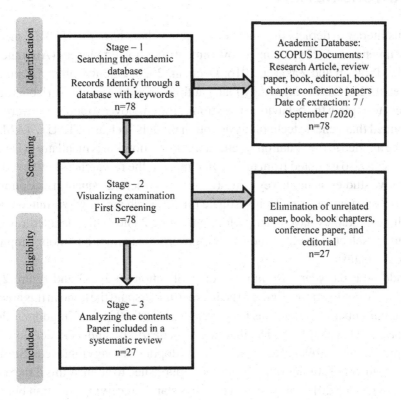

In step 3, the content was analyzed of selected papers by exploring the purpose, research methodology and results of the paper (Table -2). To understand the current trends and get gainful insight analysis was done based on year-wise publication, country-wise publication, year-wise citations of publication, study-wise number of citations and list of journals published.

Table 1. Type of document

Type of Document	Number
Articles	35
Book Chapter	7
Review	3
Conference Paper	27
Conference Review	5
Book	1
Grand Total	**78**

LITERATURE REVIEW

Previous literature depicts that most of the researchers have the UTAUT model or its sub-dimensions for exploring e-government adoption employees (Alibraheem et al., 2019; Ibrahim and Zakaria, 2016; Dečman, 2020; Olatubosun and Rao, 2012; Alraja et al., 2016). It is argued by many authors that the existing variables are not sufficient to explain the behavioral intention of the employee to adopt e-government. It is claimed that generic technology adoption models such as (UTAUT, TAM, etc.) cannot be applied to the situation where technology adoption is mandatory (Dečman, 2020). It can also be noted from table 2 that most of the researchers have conducted quantitative studies using surveys, questionnaires, structural equation modeling, and the partial squares method. To investigate this research area a very limited number of qualitative studies have been performed. Table 2 shows the detailed review of literature of selected (n=27) studies for e-government adoption among employees in various contexts.

In addition, the year-wise number of publications (table 3 and figure 2) and country-wise publications (figure 3) is shown to understand the trend in this area. The maximum number of studies has been conducted in the United Kingdom followed by Malaysia. This clearly depicts that there is a dearth of studies in developed and developing countries related to e-government adoption among employees. Similarly, a limited number of studies (n=27) have been published to address this issue. Figure 2 shows that since 2017 this research area has started receiving attention but needs to be explored more in the near future. From Table 2, it can be also interpreted that most of the researchers have explored the factors that influence employees to adopt e-government. Researchers need to explore this area from various perspectives to make successful adoption of e-government among employees.

Table 4 and Figure 4 show the citation analysis. The citation index is not very impressive; the reasoning can be less number of studies conducted in addressing employee's adoption of e-government. Only a few researchers' studies have received some good citations. Thus, it can be concluded that there a huge gap in the research literature to address this issue which needs attention. Figure 5 shows the list of Journals who have published articles on e-government related researches.

RESULTS AND DISCUSSION

The literature on e-government adoption among employees is slightly fragmented and tumbled. Researchers have investigated factors that influence the adoption of e-government by employees in recent years. Findings and results are discussed in four parts based on the selected researches. Firstly, barriers to e-government adoption

Table 2. Review of literature of selected studies

Author(s)	Objective	Research Methodology	Findings and Results
Alibraheem M.H., Abdul-Jabbar H., Ibrahim I. (2019)	To study the effect of (UTAUT) determinant on employee adoption of e-government(Electronic Tax Filing)	Quantitative research (Questionnaire survey, Partial Least Squares Method)	Performance expectancy and facilitating conditions affect the adoption of electronic tax filing by employees
Sanmukhiya C. (2019)	To address the effects of demographic factors on e-government adoption	Quantitative research (Logistic regression)	Age and education affect the decision to adopt e-government whereas marital status and gender doesn't have any effect
Al-Mutairi A., Naser K., Fayez F. (2018)	The present study examines potential obstacles in promoting e-government in Kuwait	Quantitative research (Descriptive statistics, Mann Whitney)	Kuwait has financial resources but it lacks a secured IT infrastructure, employee's low of knowledge, employee's experience in working with IT, and citizens data and information is not secured that restrict employees to adopt e-government system
Dukić D., Dukić G., Bertović N. (2017)	To assess the level of ICT literacy and background variables on employees' acceptance of e-government	Quantitative research (Questionnaire, Survey)	Employees possess basic ICT literacy and their skills level is associated with socio-demographic factors. Employees are accepting e-government but satisfaction is low among the employees with the e-government implementation process
Gupta K.P., Bhaskar P., Singh S. (2017)	To identify the factors affecting the e-government adoption by employees	Analytic hierarchy process	Two most critical aspects inducing government employees for adopting e-government are organizational and technological factors.
Al-Refaie A., Ramadna A., Bata N. (2017)	To examines the effects of various type of barriers on e-government adoption	Quantitative research (Structural equation modeling)	In successful e-government adoption, technology barriers were the key obstacle, and policy barriers, organizational barriers, strategy barriers, and end-user barriers, differed between different organizations.
Batara E., Nurmandi A., Warsito T., Pribadi U. (2017)	To examine the perspective of employees on technology acceptance variables and intention to adopt e-government	Quantitative research (Survey, Questionnaire, Structural equation modeling)	Employee's attitude is a key indicator and work experience is a significant moderating variable in adopting e-government
Althonayan M., Althonayan A. (2017)	To study the effect of e-government system on stakeholders performance	Case study	System quality variables and service quality variables have been shown to have a substantial impact on the stakeholder's performance
Yang Tsai G., Kuo T., Lin L.C. (2017)	To explore the moderating impact of management maturity on quality management and user satisfaction with e-government information platform system	Quantitative research (Survey, Questionnaire)	Management maturity shows a moderating effect on service quality; medium moderating effect on system quality and low negative moderating effect on information quality.
Ibrahim O.A., Zakaria N.H. (2016)	To determine the factors increasing e-government adoption among employees in a developing nation	Quantitative research (Survey, Questionnaire, Partial least squares regression)	The findings report UTAUT factors along with website quality, employee awareness, employee computer self-efficacy, IT employees capability and training are significant factors for e-government adoption
Alraja M.N., Hammami S., Chikhi B., Fekir S. (2016)	To investigate the effect of performance expectancy and effort expectancy on the employees intention to adopt e-government	Quantitative research (Survey, Questionnaire, Multiple, regression)	Both the factors have positive influence on employee adoption of e-government
Dečman M. (2015)	To test the UTAUT model and the influence of moderating factors for e-government adoption	Quantitative research (Survey, Questionnaire)	Employees adoption of e-government is determined by Performance expectancy and social influence and moderated by age and experience
Tbaishat R., Khasawneh S., Taamneh A.M. (2015)	To study perceptions of decision-makers towards the use of e-government	Quantitative research (Survey, Questionnaire)	The participant's perception is positive towards technology but many factors restricted participants to use computers like lack of training and technical education
Alfonso C.M., Schwarz A., Roldán J.L., Sánchez-Franco M.J. (2015)	To analyze post-adoption behaviors of employees using EDMS	Quantitative (Survey, Questionnaire, Partial least squares regression)	Employees are using Electronic Document Management System (EDMS) to its full extent and it's improving the overall performance. It is mediated by routinization and infusion.
Shajari M., Ismail Z. (2014)	To design an effective e-Government services adoption model	Quantitative research (Survey, Questionnaire, Structured Equation Modeling)	Job relevant, output, quality, and image have been added to the new e-Government services adoption model

continued on following page

25

Table 2. Continued

Author(s)	Objective	Research Methodology	Findings and Results
Zolfani S.H., Sedaghat M., Rad M.D. (2014)	To identify factors from employees perspective that influence the development and diffusion of e-government service delivery	Fuzzy analytic hierarchy process	Accessibility, private sector partnerships, security, efficiency, availability, workforce capability, information exchange are prominent factors that influence development and diffusion for e-government service delivery
Alrawabdeh W., Zeglat D. (2014)	To recognize the impact of organizational factors on the implementation of e-government	Quantitative research (Survey, Questionnaire)	Culture of organization, size of an organization, strategy of an organization, support of top management, and skill of employees affect e-government implementation
Rana N.P., Dwivedi Y.K., Williams M.D. (2013)	To propose the comprehensive review of literature pertaining to e-government	Qualitative research study (Review)	Findings revealed that there is a lack of theoretical literature. Job relevance, anxiety, perceived quality, security, perceived benefits, and are some important factors for employee's adoption but their potential benefits have not been investigated in previous researches
Kamal M.M., Hackney R., Sarwar K. (2013)	To study inhibiting factors for e-Government adoption in developing countries	Quantitative & Qualitative research study	Organizational factors, strategic factors, technological factors, political factors, operational factors, stakeholders, and social structures are critical inhibiting factors for adoption of e-Government in developing nation
Sulaiman A., Jaafar N.I., Aziz N.A.A. (2012)	To identify the impelling factors among employees to use MYEPF i-akaun (e-government website)	Quantitative research (Survey, Questionnaire)	Employee's attitude, perceived ease of use, and compatibility are impelling factors to use MYEPF i-akaun website
Olatubosun O., Rao K.S.M. (2012)	To determine adoption of e-government among the civil servant by using UTAUT model	Quantitative research (Survey, Questionnaire)	The results showed age is positively related to performance expectancy and self-efficacy additionally Social influence, attitude, and self-efficacy is positively related to gender
Al-Busaidy M., Weerakkody V. (2011)	To understand the employees perspective on the factors that may affect e-government growth and implementation	Quantitative research (Survey, Questionnaire)	Accessibility, security, privacy, efficiency, confidence, trust, availability, IT worker skills and information exchange are critical factors that affect e-government growth and implementation from the government employees perspective
Stamati T., Martakos D. (2011)	To investigate the critical success factors for adopting Unified Local Government among employees	Bibliographical survey, structured-case approach	The adoption of e-government by employees depends on political factors, cultural factors, organizational factors, technical factors, and social factors
Alhussain T., Drew S. (2010)	To identify the factors affecting employees' acceptance of e-government applications	Quantitative research (Survey, Questionnaire)	Due to digital literacy, cultural gap, technological awareness, lack of trust in technology among the employees make employees' resistance to adopt e-government
Andreopoulou Z.S. (2009)	To understand the barriers faced by employees in adopting e-government for improving government services	Quantitative research (Survey, Questionnaire)	Employees agree that e-government can be effectively introduced, but they need adequate training on the use of computer network services and resources, and it will also reduce bureaucracy
Al-Busaidy M., Weerakkody V. (2009)	To recognize the influencing factors for development and diffusion of e-government service delivery from perspective of the employees	Quantitative research (Survey, Questionnaire)	IT Employees capability influence them to use e-government.
Tseng P.T.Y., Yen D.C., Hung Y.-C., Wang N.C.F. (2008)	To discover the managerial issues related to the deployment of e-Government	Interpretative approach	National policy and institutional support become the key drivers behind IT implementation and e-Government growth in the government sector.

among employees are deliberated; secondly, enablers of e-government to support employees to adopt e-government have been reported; thirdly, benefits have been reflected to demonstrate the benefits of adopting e-government among employees.

Table 3. Year-wise number of publications

Year	No. of Publications
2008	1
2009	2
2010	1
2011	2
2012	2
2013	2
2014	3
2015	3
2016	2
2017	6
2018	1
2019	2
2020	0
Grand Total	27

Fourthly, the conceptual derived structure was suggested on the basis of the review of literature and discussion.

Figure 2. Year-wise number of publications

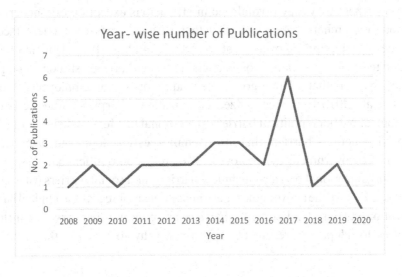

Figure 3. Country-wise number of publications

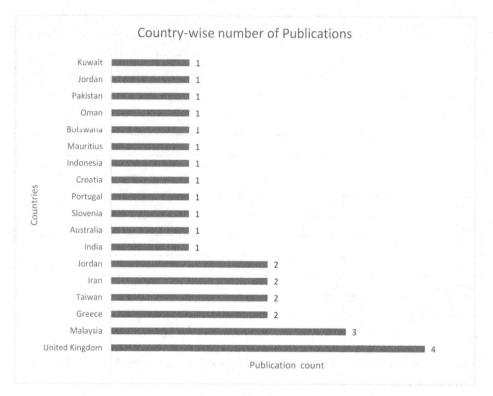

Barriers of E-Government Adoption Among Employees

Despite the substantial growth of e-government, it is not adopted successfully by employees where they play pivotal role in efficacious execution of e-government. Numerous researchers have concluded that various types of barriers faced by employees in adopting e-government. Al-Refaie et al., (2017) identified barriers related to technology, organization barriers, end-user barriers, strategy, and policy designing barriers that affect e-government adoption among employees. Similarly (Kamal et al., 2013) have recognized organizational barriers, strategic barriers, technological barriers, political barriers, operational barriers, stakeholder barriers, and social structures barriers that affect employees to adopt e-government. The adoption of e-government by employees often relies on political barriers, cultural barriers, organizational barriers, technical barriers, and social barriers (Stamati and Martakos, 2011) All the above authors have unanimously agreed that technical barriers, organization barriers, and strategic barriers are common barriers that influence employees to adopt e-government. It is noteworthy to mention that employees

Table 4. Study-wise number of citations

Research Article (Authors and Title)	Citation (Highest to Lowest)
Al-Busaidy M., Weerakkody V. (2009). E-government diffusion in Oman: a public sector employees' perspective. Transforming Government: People, Process, and Policy.	39
Tseng P.T.Y., Yen D.C., Hung Y.-C., Wang N.C.F. (2008). To explore managerial issues and their implications on e-Government deployment in the public sector: Lessons from Taiwan's Bureau of Foreign Trade. Government Information Quarterly	29
Andreopoulou, Z. S. (2009), Adoption of information and communication technologies (ICTs) in public forest service in Greece. Journal of Environmental protection and Ecology	19
Gupta, K. P., Bhaskar, P., & Singh, S. (2017). Prioritization of factors influencing employee adoption of e-government using the analytic hierarchy process. Journal of Systems and Information Technology.	13
Kamal M.M., Hackney R., Sarwar K. (2013). Investigating factors inhibiting e-government adoption in developing countries: the context of Pakistan. Journal of Global Information Management	13
Al-Busaidy, M., & Weerakkody, V. (2011). E-government services in Oman: an employee's perspective. Electronic Government, an International Journal	12
Rana, N. P., Dwivedi, Y. K., & Williams, M. D. (2013). E-government adoption research: An analysis of the employee's perspective. International Journal of Business Information Systems	10
Batara, E., Nurmandi, A., Warsito, T., & Pribadi, U. (2017). Are government employees adopting local e-government transformation?. Transforming Government: People, Process and Policy.	9
Althonayan, M., & Althonayan, A. (2017). E-government system evaluation. Transforming Government: People, Process and Policy.	5
Stamati, T., & Martakos, D. (2013). Electronic transformation of local government: An exploratory study. In E-Government Services Design, Adoption, and Evaluation (pp. 20-38). IGI Global.	5
Sulaiman, A., Jaafar, N. I., & Aziz, N. A. A. (2012). Factors influencing intention to use MYEPF I-Akaun. World Applied Sciences Journal	4
Alhussain, T., & Drew, S. (2010). Employees' perceptions of biometric technology adoption in e-government: An exploratory study in the kingdom of saudi arabia. International Journal of E-Adoption	4
Al-Refaie, A., & Ramadna, A. M. (2020). Barriers to E-Government Adoption in Jordanian Organizations from Users' and Employees' Perspectives. In Open Government: Concepts, Methodologies, Tools, and Applications	3
Ibrahim, O. A., & Zakaria, N. H. (2016). E-government services in developing countries: a success adoption model from employees perspective. Journal of Theoretical & Applied Information Technology	3
Alraja, M. N., Hammami, S., Chikhi, B., & Fekir, S. (2016). The Influence of Effort and Performance Expectancy on Employees to adopt e-Government: evidence from Oman. International Review of Management and Marketing	3
Dečman, M. (2020). Understanding Technology Acceptance of Government Information Systems from Employees' Perspective. In Open Government: Concepts, Methodologies, Tools, and Applications	3
Afonso, C. M., Schwarz, A., Roldán, J. L., & Sánchez-Franco, M. J. (2015). EDMS use in local E-government: An analysis of the path from extent of use to overall performance. International Journal of Electronic Government Research	3
Olatubosun, O., & Madhava Rao, K. S. (2012). Empirical study of the readiness of public servants on the adoption of e-government. International Journal of Information Systems and Change Management	3
Abdullah, A., Naser, K., & Fayez, F. (2018). Obstacles toward adopting electronic government in an emerging economy: Evidence from Kuwait. Asian Economic and Financial Review	2
Tsai, G. Y., Kuo, T., & Lin, L. C. (2017). The moderating effect of management maturity on the implementation of an information platform system. Journal of Organizational Change Management.	2
Hashemkhani Zolfani, S., Sedaghat, M., & Rad, M. D. (2014). E-government diffusion in Iran: a public sector employees' perspective. International Journal of Business Information Systems	2
Dukić, D., Dukić, G., & Bertović, N. (2017). Public administration employees' readiness and acceptance of e-government: Findings from a Croatian survey. Information Development	1
Alibraheem M.H., Abdul-Jabbar H., Ibrahim I. (2019) Electronic tax filing adoption in Jordan: The tax employees' perspectives. International Journal of Advanced Science and Technology	0
Sharma, S. K. (2015). Adoption of e-government services. Transforming Government: People, Process and Policy.	0
Tbaishat, R. M., & Khasawneh, S. (2015). The Decision Makers' Perception Toward The Adoption of Information Technology By Government Institutions In jordan and Its affect on InformationAccessibility, and Decision Making Quality. Journal of Social and Administrative Sciences	0
Shajari, M., & Ismail, Z. (2014). Constructing an adoption model for e-government services. Jurnal Teknologi	0
Alrawabdeh, W. (2014). The Impact of Environmental Factors on E-Government Implementation: The Case of Jordan. International Journal of Management Sciences and Business Research	0
	187

Figure 4. Year-wise citation of publications

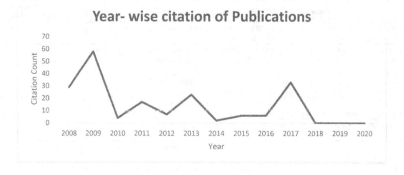

Figure 5. List of Journals with the number of articles published

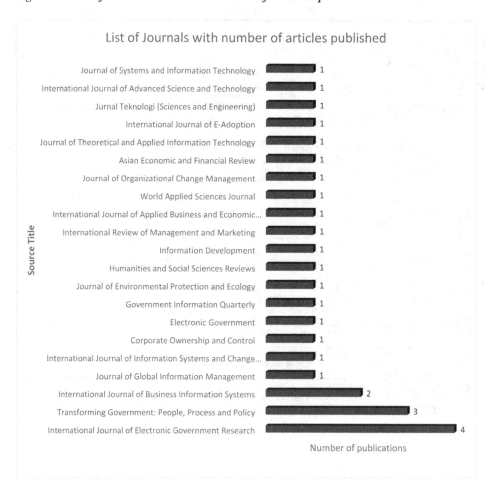

want e-government systems to be implemented but they are not confident in the e-government security feature and lack trust in protection of information and data (Abdullah et al., 2018; Al-Busaidy and Weerakkody 2011; Al-Mutairi et al., 2018; Dukić et al., 2017). Alhussain and Drew (2010) indicate digital literacy and the cultural gap in technological development among the employees. The organization doesn't have capable IT employees, as they lack awareness, lack computer self-efficacy that need to eradicate by providing education and training on operating e-government (Ibrahim and Zakaria, 2016; Tbaishat et al., 2017). It has been highlighted by many researchers that employees are not well trained to operate e-government besides lack of technical support they find it more difficult to work on e-government (Gupta et al., 2017). The organization does not provide sufficient technical infrastructure to work and poor internet connectivity makes employees resist adopting e-government (Gupta et al., 2017). Employees require proper hardware and software to work on the e-government website. E-government should be flexible, compatible, and easy to use with the availability of correct data at right time for the employees (Sulaiman et al., 2012). Althonayan and Althonayan, 2017) examined system quality factors and service quality factors for e-government adoption and improving performance. Ibrahim and Zakaria, 2016) advocate that poor website quality irritates employees for working on e-government. Government websites are poorly designed and lack visual appeals.

Enablers of E-Government Adoption Among Employees

Although there are several barriers faced by employees in adopting e-government but enablers facilitates employees to adopt e-government. The government should focus on enablers to facilitate progression for e-government adoption among employees. The majority of the researchers have reported that the organization plays a vital role as an enabler in making successful e-government adoption among employees. The government organization needs to provide necessary facilitating conditions, technical support, legal support, and financial support for implementing e-government and adaptable in an efficacious manner among employees (Batara et al., 2017). Organizations need to create a positive culture through a robust organization strategy for implementing e-government (Alrawabdeh and Zeglat, 2014). The barriers faced by employees can be mitigated with the support of organizations. Like, an organization can allocate specific funds for purchasing specific technological resources for e-government development, funds for providing training to the employees from external experts. Tseng et al., (2017) suggested that the national policy for e-Government implementation is also a critical enabler for adoption of e-government among employees. A clear e-government roadmap can help in successfully implementing e-government in any organization. Due to a lot of ambiguities during implementation, employees may misinterpret the reason for making changes from traditional government to e-government. Employees perceive e-government has increased task complexities and workload. The government needs to make put effort into maintaining a positive attitude among employees toward e-government (Batara et al., 2017). This can be done by providing adequate information about the e-government, benefits of e-government, and challenges that may be likely to be faced by them. The organization can assure employees that they are available for their support by building a strong psychological contract. It has also been observed that employees lack trust in technology that needs to be narrow down with help of top management support (Alhussain and Drew, 2010). If employees will develop trust in e-government, they will also convince the citizen to use e-government (Al-Busaidy and Weerakkody, 2011).

Additional, many countries such as Egypt (Soliman et al.,2020); China et al., 2020); Jordan (Al-Shqairat et al., 2014); Malaysia (Kaliannan et al., 2010); Poland (Ziemba et al., 2013); Singapore (Taher et al., 2012) are operating e-government thorough Public-private partnership model. This is also a significant enabler for e-government adoption among employees. The government can join hands with the private sector by outsourcing some work/tasks based on the requirements. This will allow employees to focus on the important work area which requires their attention. Public-private partnerships can enhance the efficiency of E-government adoption among employees (Zolfani et al., 2014). Furthermore, the Government should consider Socio-demographic factors while recruiting employees. Socio-demographic factors can be an important enabler, as most of the previous studies have proven that employees education (Sanmukhiya, 2019), computers skill (Dukić et al., 2017; Alrawabdeh and Zeglat, 2014), age (Olatubosun and Rao, 2012) are critical factors influencing employees to adopt e-government. The employee who possesses high education levels and computer skills show a positive attitude towards technology adoption (Quazi & Talukder, 2011). Also, old age employees tend to be old-fashioned, also like to stick to traditional working patterns. Young age employees are more tech-savvy compared to their counterparts. They don't face many difficulties in working with e-government (Elias et al., 2012). The government should deploy young employees with high education and computer skills to work on e-government for the effective implementation of e-government.

Benefits of E-Government Adoption Among Employees

While reviewing the selected literature, the benefits of adopting e-government among employees have not been explored. Very few researchers have touched this area, the reason can be attributed that there are more barriers than benefits in adopting e-government among employees. Employees have not accepted e-government truly and they are still dealing with barriers so it becomes quite obvious that benefits cannot be realized until it is successfully adopted by employees. Many employees have negative opinions towards e-government and are not willing to adopt because they do not explicitly see many benefits directly from e-government. User acceptance of technology is a significant element for technology adoption (Alibraheem et al., 2019). The benefits of e-government are mainly citizen-oriented where they enjoy better service at a minimum cost in less time (Almaiah & Nasereddin, 2020; Li & Shang, 2020; Gupta et al., 2015). The efficient delivery of services is possible through employees so they need to work on the e-government system mandatory. Employees are forced to adopt e-government without giving them adequate information and knowledge of e-government. Previous studies have proven that e-government helps employees in reducing efforts and improve their job performance (Rehouma & Hofmann, 2018; Jeloudarlu and Begzadeh, 2016). Performance expectancy and effort expectancy are two major perceived benefits of e-government adoption among employees (Alibraheem et al., 2019; Alraja et al., 2016; Olatubosun and Rao, 2012). It's an employee tendency that they feel motivated to work when they get some benefits from it (Govindarajulu & Daily, 2004; Burton, (2012). The government needs to support employees to realize e-government the benefits fully. Selected studies have reported that e-government has helped employees upgrading their ICT skills (Al-Busaidy & Weerakkody, 2009). Employees use this IT skill in dealing with their personal work as well like payment of the bill, online booking, complaint filing, etc. This makes employees more acceptable in this tech society where it has become important to be tech-savvy to remain in trend. Many previous researchers have found social influence as important dimensions of e-government adoption (Shajari and Ismail, 2014; Olatubosun and Rao, 2012; Batara et al., 2017) Dečman, 2015; Alraja et al., 2016). Due to the dearth of literature on the benefits of adopting e-government, future research needs to identify the various categories of benefits for adopting e-government among employees.

Proposed Systematic Review-Derived Conceptual Framework

The conceptual derived framework has been suggested on the basis of systematic review and discussion. The proposed framework highlights the barriers, enablers, and benefits of adopting e-government among employees (Table 5 & Figure 6). The major barriers faced by employees are lack of capable IT employees, cultural barriers, digital literacy, end-user barriers, lack of awareness, service quality factors, lack of computer self-efficacy, social barriers, lack of technical support, security feature, lack of training, lack of trust, operational barriers, organizational barriers, policy barriers, political barriers, poor internet connectivity, social structures barriers, stakeholder's barriers, strategic barriers, system quality factors, technological barriers, and website quality. The government needs to focus on removing these barriers to make employees adopt e-government. There are many enablers that help in promoting e-government adoption among employees creating awareness. These enablers are facilitating conditions, financial support, legal support, national policy, organization culture, organization strategy, public-private partnership model, socio-demographic factors, technical support staff, and top management support. If the government can successfully remove the barriers, the benefits of adopting e-government among employees can be realized in the real terms. In the current study, the benefits of adopting e-government among employees are reducing efforts, improve job performance, upgrade ICT skills, and improves image in society. Most of the previous studies have discussed the barriers in adopting e-government among the employees and few of also explored the enablers of adopting e-government. In this research, the authors propose additional benefits that may be reaped by employees if they adopt e-government adoption successfully. The proposed benefits are suggested on the basis of technology adoption by employees in others sectors like banking (Imran, et al., 2014) beverage companies (Tusiimemukama, 2019), non-governmental organizations (Al-Nashmi and Amer, (2014), manufacturing companies (Boothby, et al., 2010). Technology adoption has created better communication and coordination among employees (Mirvis, et al., 1991), enhanced task compliance (Darwazeh, et

Figure 6. A proposed systematic review-derived conceptual framework

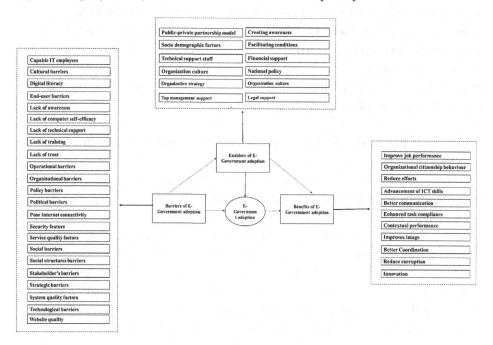

Table 5. A proposed systematic review-derived conceptual framework

E-Government Adoption Among Employees		
Barriers	**Enablers**	**Benefits**
• Operational barriers • Technological barriers • Lack of awareness • Policy barriers • Capable IT employees • Cultural barriers • Lack of computer Digital literacy • End-user barriers • Lack of training • Political barriers • Lack of technical support • Poor internet connectivity • Security feature • Service quality factors • Lack of self-efficacy • Social barriers • Stakeholder's barriers • Strategic barriers • Lack of trust • System quality factors • Website quality	• Public-private partnership model • Top management support • National policy • Technical support staff • Organization culture • Creating awareness • Facilitating conditions • Financial support • Legal support • Socio-demographic factors • Organization strategy	• Improve job performance • Improves image • Reducing efforts • Advancement of Information and Communication Technologies (ICT) skills *Proposed benefits:* • Organizational citizenship behavior • Better communication and Coordination • Contextual performance • Enhanced task compliance • Innovation • Reduce corruption

al., 2016), reduce corruption (Bhattacherjee & Shrivastava, U. (2018) contextual performance (Peng et al., 2016), organizational citizenship behavior (Luciano et al., 2016) and innovation Chen & Tsou, 2007).

CONCLUSION

The research has attempted to conduct a systematic literature review to identify barriers, enablers and benefits of adopting e-government among employees and recommending a systematic review-derived conceptual framework. The systematic review of literature clearly depicts that there is a dearth of literature in this area, which needs to be addressed at the earliest by the researchers to make e-government adoption successful among employees. On the basis of the systematic review of literature it can be concluded that most of the studies have been conducted to explore barriers of e-government adoption among employees. Very limited research has been conducted on enablers and the benefits of e-government adoption among employees. This research has suggested some additional benefits that can be derived when employees adopt e-government successfully. This research presents the proposed systematic review and derives conceptual framework that holistically combines barriers, enablers and benefits of E-government adoption among employees. The validity and reliability of the proposed conceptual model need to be empirically tested in future research studies. The contribution of additional proposed benefits for adopting e-government also needs to be analyzed in future.

REFERENCES

Abdullah, A., Naser, K., & Fayez, F. (2018). Obstacles toward adopting electronic government in an emerging economy: Evidence from Kuwait. *Asian Economic and Financial Review*, *8*(6), 832–842. doi:10.18488/journal.aefr.2018.86.832.842

Afonso, C. M., Schwarz, A., Roldán, J. L., & Sánchez-Franco, M. J. (2015). EDMS use in local E-government: An analysis of the path from extent of use to overall performance. *International Journal of Electronic Government Research*, *11*(2), 18–34. doi:10.4018/IJEGR.2015040102

Al-Busaidy, M., & Weerakkody, V. (2009). E-government diffusion in Oman: a public sector employees' perspective. Transforming Government: People. *Process and Policy*, *3*(4), 375–393.

Al-Busaidy, M., & Weerakkody, V. (2011). E-government services in Oman: an employee's perspective. *Electronic Government, an International Journal, 8*(2-3), 185-207.

Al-Nashmi, M. M., & Amer, R. A. (2014). The Impact of Information Technology Adoption on Employee Productivity In Nongovernemental Organizations In Yemen. *International Journal of Social Sciences and Humanities Research*, *3*(2), 32–50.

Al-Refaie, A., & Ramadna, A. M. (2020). Barriers to E-Government Adoption in Jordanian Organizations from Users' and Employees' Perspectives. In Open Government: Concepts, Methodologies, Tools, and Applications (pp. 2190-2210). IGI Global.

Al-Shqairat, Z. I., Al Shra'ah, A. E., Al-rawad, M. I., & Al-Kilani, D. H. (2014). Assessing the planning of Public Private Partnership (PPP) in E-government implementation experience in Jordan. *International Journal of Business and Management*, *9*(2), 124. doi:10.5539/ijbm.v9n2p124

Alhussain, T., & Drew, S. (2010). Employees' perceptions of biometric technology adoption in e-government: An exploratory study in the kingdom of Saudi Arabia. *International Journal of E-Adoption*, *2*(1), 59–71. doi:10.4018/jea.2010010105

Alibraheem, M. H., Abdul-Jabbar, H., & Ibrahim, I. (2019). Electronic tax filing adoption in Jordan: The tax employees' perspectives. *International Journal of Advanced Science and Technology, 28*(17), 681–689.

Almaiah, M. A., & Nasereddin, Y. (2020). Factors influencing the adoption of e-government services among Jordanian citizens. Electronic Government. *International Journal (Toronto, Ont.), 16*(3), 236–259.

Alraja, M. N. (2016). The effect of social influence and facilitating conditions on e-government acceptance from the individual employees' perspective. *Polish Journal of Management Studies, 14.*

Alraja, M. N., Hammami, S., Chikhi, B., & Fekir, S. (2016). The Influence of Effort and Performance Expectancy on Employees to adopt e-Government: Evidence from Oman. *International Review of Management and Marketing, 6*(4).

Alrawabdeh, W. (2014). The Impact of Environmental Factors on E-Government Implementation: The Case of Jordan. *International Journal of Management Sciences and Business Research, 3*(3).

Althonayan, M., & Althonayan, A. (2017). E-government system evaluation: The case of users' performance using ERP systems in higher education. Transforming Government: People. *Process and Policy, 11*(3), 306–342.

Alzahrani, L., Al-Karaghouli, W., & Weerakkody, V. (2017). Analysing the critical factors influencing trust in e-government adoption from citizens' perspective: A systematic review and a conceptual framework. *International Business Review, 26*(1), 164–175. doi:10.1016/j.ibusrev.2016.06.004

Andreopoulou, Z. S. (2009). Adoption of information and communication technologies (ICTs) in public forest service in Greece. *Journal of Environmental Protection and Ecology, 10*(4), 1194–1204.

Azim, R. M. H. A., Salman, O., & El Henawy, I. (2020). The Role of E-Government as a Stimulus for Economic Growth. *The International Journal of Business Management and Technology, 4*(5), 69–79.

Batara, E., Nurmandi, A., Warsito, T., & Pribadi, U. (2017). Are government employees adopting local e-government transformation? Transforming Government: People. *Process and Policy, 11*(1), 612–638.

Batubara, F. R., Ubacht, J., & Janssen, M. (2018, May). Challenges of blockchain technology adoption for e-government: a systematic literature review. In *Proceedings of the 19th Annual International Conference on Digital Government Research: Governance in the Data Age* (pp. 1-9). 10.1145/3209281.3209317

Bhattacherjee, A., & Shrivastava, U. (2018). The effects of ICT use and ICT Laws on corruption: A general deterrence theory perspective. *Government Information Quarterly*, *35*(4), 703–712. doi:10.1016/j.giq.2018.07.006

Boothby, D., Dufour, A., & Tang, J. (2010). Technology adoption, training and productivity performance. *Research Policy*, *39*(5), 650–661. doi:10.1016/j.respol.2010.02.011

Burton, K. (2012). A study of motivation: How to get your employees moving. *Management*, *3*(2), 232–234.

Chen, J. S., & Tsou, H. T. (2007). Information technology adoption for service innovation practices and competitive advantage: The case of financial firms. *Information Research*, *12*(3), n3.

Darwazeh, R., Khraisat, D., & Al Dajah, S. (2016). The effect of application of e-government on the staff performance in the Greater Amman Municipality a field study. *Research in Business and Management*, *3*(2), 19–40. doi:10.5296/rbm.v3i2.9640

Dečman, M. (2020). Understanding Technology Acceptance of Government Information Systems from Employees' Perspective. In Open Government: Concepts, Methodologies, Tools, and Applications (pp. 1488-1507). IGI Global.

Dukić, D., Dukić, G., & Bertović, N. (2017). Public administration employees' readiness and acceptance of e-government: Findings from a Croatian survey. *Information Development*, *33*(5), 525–539. doi:10.1177/0266666916671773

Elias, S. M., Smith, W. L., & Barney, C. E. (2012). Age as a moderator of attitude towards technology in the workplace: Work motivation and overall job satisfaction. *Behaviour & Information Technology*, *31*(5), 453–467. doi:10.1080/0144929X.2010.513419

Franke, R., & Eckhardt, A. (2014). *Crucial Factors for E Government Implementation Success and Failure: Case Study Evidence from Saudi Arabia*. Academic Press.

Gilbert, D., Balestrini, P., & Littleboy, D. (2004). Barriers and benefits in the adoption of e-government. *International Journal of Public Sector Management*, *17*(4), 286–301. doi:10.1108/09513550410539794

Govindarajulu, N., & Daily, B. F. (2004). Motivating employees for environmental improvement. *Industrial Management & Data Systems, 104*(4), 364–372. doi:10.1108/02635570410530775

Gupta, K. P., Bhaskar, P., & Singh, S. (2015). A Conceptual Framework for Measuring Benefits of E-Governance. *Public Policy and Administration Research, 5*(12), 30–35.

Gupta, K. P., Bhaskar, P., & Singh, S. (2017). Prioritization of factors influencing employee adoption of e-government using the analytic hierarchy process. *Journal of Systems and Information Technology, 19*(1/2), 116–137. doi:10.1108/JSIT-04-2017-0028

Hashemkhani Zolfani, S., Sedaghat, M., & Rad, M. D. (2014). E-government diffusion in Iran: A public sector employees' perspective. *International Journal of Business Information Systems, 15*(2), 205–221. doi:10.1504/IJBIS.2014.059251

Ibrahim, O. A., & Zakaria, N. H. (2016). E-government services in developing countries: A success adoption model from employees perspective. *Journal of Theoretical and Applied Information Technology, 94*(2).

Ingrams, A., Manoharan, A., Schmidthuber, L., & Holzer, M. (2018). Stages and determinants of e-government development: A twelve-year longitudinal study of global cities. *International Public Management Journal,,* 1–39. doi:10.1080/1096 7494.2018.1467987

Ismail, I., Fathonih, A., Prabowo, H., Hartati, S., & Redjeki, F. (2020). Transparency and Corruption: Does E-Government Effective to Combat Corruption? *International Journal of Psychosocial Rehabilitation, 24*(4), 5396–5404. doi:10.37200/IJPR/ V24I4/PR201636

Kaliannan, M., Awang, H., & Raman, M. (2010). Public-private partnerships for e-government services: lessons from Malaysia. *Institutions and Economies*, 207-220.

Kamal, M. M., Hackney, R., & Sarwar, K. (2013). Investigating factors inhibiting e-government adoption in developing countries: The context of Pakistan. *Journal of Global Information Management, 21*(4), 77–102. doi:10.4018/jgim.2013100105

Kettani, D. (2014). Technology Enablers for E-Government Systems. In E-Government for Good Governance in Developing Countries: Empirical Evidence from the eFez Project (pp. 223-250). Anthem Press.

Li, Y., & Shang, H. (2020). Service quality, perceived value, and citizens' continuous-use intention regarding e-government: Empirical evidence from China. *Information & Management, 57*(3), 103197. doi:10.1016/j.im.2019.103197

Liu, T., Yang, X., & Zheng, Y. (2020). Understanding the evolution of public–private partnerships in Chinese e-government: Four stages of development. *Asia Pacific Journal of Public Administration,*, 1–26.

Luciano, E. M., Wiedenhöft, G. C., Macadar, M. A., & dos Santos, F. P. (2016). Information technology governance adoption: understanding its expectations through the lens of organizational citizenship. *International Journal of IT/Business Alignment and Governance (IJITBAG), 7*(2), 22-32.

Mahmood, M., Osmani, M., & Sivarajah, U. (2014). *The role of trust in e-government adoption: A systematic literature review.* Academic Press.

Mirvis, P. H., Sales, A. L., & Hackett, E. J. (1991). The implementation and adoption of new technology in organizations: The impact on work, people, and culture. *Human Resource Management, 30*(1), 113–139. doi:10.1002/hrm.3930300107

Mustafa, A., Ibrahim, O., & Mohammed, F. (2020). E-government adoption: A systematic review in the context of developing nations. *International Journal of Innovation, 8*(1), 59–76. doi:10.5585/iji.v8i1.16479

Olatubosun, O., & Madhava Rao, K. S. (2012). Empirical study of the readiness of public servants on the adoption of e-government. *International Journal of Information Systems and Change Management, 6*(1), 17–37. doi:10.1504/IJISCM.2012.050337

Othman, M. H., Razali, R., & Faidzul, M. (2020). Key Factors for E-Government towards Sustainable Development Goals. *Development, 29*(6s), 2864-2876.

Peng, J., Zhang, G., & Dubinsky, A. J. (2016). Knowledge sharing, social relationships, and contextual performance: The moderating influence of information technology competence. Business intelligence: Concepts, methodologies, tools, and applications, 491-1506.

Quazi, A., & Talukder, M. (2011). Demographic determinants of adoption of technological innovation. *Journal of Computer Information Systems, 52*(1), 34–42.

Rana, N. P., Dwivedi, Y. K., & Williams, M. D. (2013). E-government adoption research: An analysis of the employee's perspective. *International Journal of Business Information Systems, 14*(4), 414–428. doi:10.1504/IJBIS.2013.057497

Rana, N. P., Dwivedi, Y. K., & Williams, M. D. (2013). E-government adoption research: An analysis of the employee's perspective. *International Journal of Business Information Systems, 14*(4), 414–428. doi:10.1504/IJBIS.2013.057497

Rehouma, M. B. (2020). Exploring the Role of Participation in Government Employees' Adoption of IT: A Qualitative Study of Employees' Participation in the Introduction of the E-File in Germany. *International Journal of Public Administration in the Digital Age, 7*(1), 33–46. doi:10.4018/IJPADA.2020010103

Rowley, J. (2011). E-Government stakeholders—Who are they and what do they want? *International Journal of Information Management, 31*(1), 53–62. doi:10.1016/j.ijinfomgt.2010.05.005

Shajari, M., & Ismail, Z. (2014). Constructing an adoption model for e-government services. *Jurnal Teknologi, 68*(2). Advance online publication. doi:10.11113/jt.v68.2906

Sharma, S. K. (2015). Adoption of e-government services: The role of service quality dimensions and demographic variables Transforming Government: People. *Process and Policy, 9*(2), 207–2022.

Soliman, K., & El-Barkouky, N. (2020, May). An E-government Procurement Decision Support System Model for Public Private Partnership Projects in Egypt. In *International Conference on Decision Support System Technology*. Springer. 10.1007/978-3-030-46224-6_9

Srivastava, S. C., & Teo, T. S. (2004, July). A framework for electronic government: evolution, enablers and resource drainers. *Proceedings of the Eighth Pacific Asia Conference on Information Systems*.

Stamati, T., & Martakos, D. (2013). Electronic transformation of local government: An exploratory study. In E-Government Services Design, Adoption, and Evaluation (pp. 20-38). IGI Global.

Stefanovic, D., Marjanovic, U., Delić, M., Culibrk, D., & Lalic, B. (2016). Assessing the success of e-government systems: An employee perspective. *Information & Management, 53*(6), 717–726. doi:10.1016/j.im.2016.02.007

Sulaiman, A., Jaafar, N. I., & Aziz, N. A. A. (2012). Factors influencing intention to use MYEPF I-Akaun. *World Applied Sciences Journal, 18*(3), 451–461.

Suzuki, T., & Suzuki, L. (2020). On the benefit of 3-tier SOA architecture promoting information sharing among TMS systems and Brazilian e-Government Web Services: A CT-e case study. arXiv preprint arXiv:2005.13047.

Taher, M., Yang, Z., & Kankanhalli, A. (2012, July). Public-Private Partnerships In E Government: Insights From Singapore Cases. In PACIS (p. 116). Academic Press.

Tsai, G. Y., Kuo, T., & Lin, L. C. (2017). The moderating effect of management maturity on the implementation of an information platform system. *Journal of Organizational Change Management, 30*(7), 1093–1108. doi:10.1108/JOCM-03-2017-0048

Tseng, P. T., Yen, D. C., Hung, Y. C., & Wang, N. C. (2008). To explore managerial issues and their implications on e-Government deployment in the public sector: Lessons from Taiwan's Bureau of Foreign Trade. *Government Information Quarterly, 25*(4), 734–756. doi:10.1016/j.giq.2007.06.003

Tusiimemukama, A. (2019). *Technology adoption employee engagement and perceived job performance in Uganda beverage companies* (Doctoral dissertation). Kyambogo University.

United Nations. (2003). *World Public Sector Report 2003: E-Government at the Crossroads.* United Nations. Available at: https://publicadministration.un.org/publications/content/PDFs/ELibrary%20Archives/World%20Public%20Sector%20Report%20series/World%20Public%20Sector%20Report.2003.pdf

United Nations. (2015). *Transforming Our World: The 2030 Agenda for Sustainable Development, General Assembly.* United Nations. Available at: https://undocs.org/A/RES/70/1

United Nations. (2018). *United Nations E-Government Survey 2018: Gearing e-Government to Support Transformation towards Sustainable and Resilient Societies.* United Nations.

United Nations Department of Economic and Social Affairs (UNDESA). (2003). *World Public Sector Report: E-Government at the Crossroads.* United Nations.

Wang, S., & Feeney, M. K. (2016). Determinants of information and communication technology adoption in municipalities. *American Review of Public Administration, 46*(3), 292–313. doi:10.1177/0275074014553462

Zarei, B., Saghafi, F., Zarrin, L., & Ghapanchi, A. H. (2014). An e-government capability model for government employees. *International Journal of Business Information Systems, 16*(2), 154–176. doi:10.1504/IJBIS.2014.062836

Zhang, J., Dawes, S. S., & Sarkis, J. (2005). Exploring stakeholders' expectations of the benefits and barriers of e-government knowledge sharing. *Journal of Enterprise Information Management, 18*(5), 548–567. doi:10.1108/17410390510624007

Ziemba, E. (2018). The ICT adoption in government units in the context of the sustainable information society. In *2018 Federated Conference on Computer Science and Information Systems (FedCSIS)* (pp. 725-733). IEEE. 10.15439/2018F116

Ziemba, E., Papaj, T., & Żelazny, R. (2013). A model of success factors for e-government adoption-the case of Poland. *Issues in Information Systems*, *14*(2).

Chapter 3
Innovations in HRM Practices in Indian Companies:
A Review–Based Study

Minisha Gupta
Quality Cognition Private Limited, India

ABSTRACT

Innovation has become an integral part of every business organization because it provides sustainable competitive advantage to the company. In today's highly dynamic business environment every organization wants to succeed by leveraging their employees' talent. In order to leverage the employees' talent, it is important to follow relevant HRM policies or to continuously introduce innovative HR practices to meet the expectations of the employees. Reviewing the literature and previous research work, this study has tried to find out the various innovative HR practices initiated by Indian organizations till now. The findings will help in guiding how much more innovative practices are still to be initiated in order to attain employee confidence and loyalty for the company. The study is beneficial for business leaders, students, practitioners, and researchers.

INTRODUCTION

Organizations are facing major challenges resultant of globally dynamic and uncertain business environment, thrive to complete consumers' demand and meet their expectations, rapidly changing technology and enhanced work structures. This constant pressure to sustain in the competition and attain competitive edge over others is forcing organizations to leverage their talent. However, with the

DOI: 10.4018/978-1-7998-4180-7.ch003

changing expectations of their talented employees, it has become more difficult for companies to retain the talent and the only key for this is innovation in HRM policies (Gupta, 2018). Although, with the technological advancements, most of the companies have experimented with some innovations in their HRM policies like initiating Sustainable HRM (Miles & Snow, 1984; Andrade & Lengnick-Hall et al., 2009; Kramar, 2014) or Green HRM (Renwick, Redman, & Maguire, 2008; Mandip, 2012; Muster & Schrader, 2011), yet, more innovative HRM policies needs to be initiated for retaining talented workforce and attaining sustainable competitive advantage. Thus, this study has been taken into account to identify the innovative HRM policies implemented in the Indian companies and what else needs to be initiated by companies to attain sustainable competitive advantage. The study reviews previous literature and information available on relevant HR websites to gather the significant data for this research work.

BACKGROUND

Human Resource Management or HRM is an integral and most important function for business organizations. HR managers are not only responsible to hire relevant talent for the companies but also to ensure their constant productive output for the organizational growth and success. They also take care of the training and development facilities to nurture the talent of the employees and ask employees to give their feedback or suggestions on existing policies for further improvement. However, talent attrition and organizational sustainability are still two important challenges in front of business managers (Naim & Lenka, 2017; Lenka, Gupta, & Sahoo, 2016). With the constant research work, many practitioners and researchers have suggested to change old methods and policies of HRM and offered a variety of creative and innovative HR policies to improve organizational productivity. Creativity infuses novel and useful ideas or solutions to the problems of the company and innovation helps in implementing those ideas in organizational settings (Lenka, Gupta, & Sahoo, 2016; Gupta, 2018). As, employees are the main focus point of managers while developing HR policies, thus any creative and innovative HR policy can only be completed when it includes employees' welfare. However, previous research work shows little evidence suggesting that innovation is focussed on other functions of companies rather than focusing on HR policies (Shipton, Sparrow, Budhwar, & Brown, 2017b). Research questions in HRM have mainly take into consideration the role of HRM policies or framework in organizational or employee performance in different sectors across the world including every type of organization whether public or private, small or large (Boxall & Purcell, 2011; Boxall & Purcell, 2016; Cavanagh et al., 2017; Cooke & Saini, 2010; Datta et al., 2005; Stanton et al.,

2014;). However, with the new research paradigms, focus of researchers shifted from organizational performance to organizational employees in terms of commitment, retention, engagement, and well being (Bamber et al., 2017; Boxall & Purcell, 2016; Guest & Conway, 2011; Ulrich, 2016). This research is in nascent stage of linking the role of HRM policies in employee creativity and innovation leading to organizational sustainability through creative, virtual, or Research and Development (R&D) teams (Lenka & Gupta, 2019; Lenka, Gupta, & Sahoo, 2016; Shipton et al., 2017a). Therefore, it is important to explore this field and find out the creative and innovative HRM policies leading to growth and sustainability of organizations.

Traditionally, innovation is considered as a scientific or technological method or procedure to improve the products, process, or structures of the organizations in order to achieve organizational sustainability (Gupta, 2018). Innovation is considered as a means to initiate and implement creative products and processes in the organization for organizational growth and sustainability (Damanpour, 1991; Rogers, 1983). Innovative HR practices are defined as ideas, policies, programmes, or systems adopted by organizations to enhance their employees' performance (Wolfe, 1995; Agarwala, 2003). Nevertheless, innovation is not now limited to products, process, or organizational structures. Researchers, practitioners, and business managers are now considering innovation into HRM also in order to frame creative and innovative HRM policies for the better productivity and commitment of employees towards their organization. Strategic HRM, Green HRM, E-HRM, are few innovative HRM policies implemented in organizations to improve the existing HR processes and to raise employees' environmental awareness at their workplace simultaneously (Collins, 2020; Mishra, Sarkar & Kiranmai, 2014; Renwick, Redman, & Maguire, 2008; Ruël, Bondarouk, & Looise, 2004; Shafaei, Nejati, & Yusoff, 2020). Intrapreneurship, where employees behave like an entrepreneur for the product or idea they brought to the company, is also an innovative HR policy initiated by organizations to enhance creativity and innovation for their growth (Antoncic & Hisrich, 2001; Antoncic, 2007; Amado et al., 2009; Gupta, 2016; Parker, 2011).

Apart from the above defined innovations in HR, it has become necessary for organizations to implement new and innovative HR policies related to intrinsic rewards, job satisfaction, accomplishing organizational goals, performance and promotions, participation in brainstorming and decision making sessions, and training and development. A recent research work has been carried out on the HRM in Indian organizations by conducting a survey on 252 HR directors which revealed that, Strategic HRM is having limitations in bridging the gap between expected and original outcomes of firm (Darwish et al., 2019). Indian organizations are opening up to implement innovative HR practices but they need to maintain systemic support and strong infrastructural background before implementing innovative HR practices. A study conducted on Indian hotel industry by surveying 52 HR managers, 260 customers and employees revealed that, innovation in HRM practices improve hotel service effectiveness resulting in enhanced organizational performance (Chand, 2010). Similarly, IT, banking, automobile, manufacturing, service, and healthcare industries in India are also implementing HPWS (High performance work systems) in their HR functions (Malik et al., 2017; Srinivasan & Chandwani, 2014; Kong, Chadee, & Raman, 2012; Mishra, Sarkar, & Kiranmai, 2014).

In recent years, India has become the hub of innovation and entrepreneurship which accelerated India's economic growth by attracting more foreign investments in the form of joint ventures and partnership firms (Fan, 2011; Kulkarni, 2013). To maintain this pace of growth and to sustain in global and domestic markets, it has become important for Indian organizations to continuously create and innovate by leveraging the talent of their employees. With the changes in technology and society, the expectations of the employees especially Generation Y need to be met in order to retain them in the organization. These Generation Y employees are the technological literates who believe in maintaining work life balance and look for meaningful tasks (Naim, 2014). They want to seek attention and achievement for their task along with technological advancements and free access to organizational resources. To fulfil their requirements and to engage them to work for organizational creativity and innovation, it has become must for organizations to remove their old and redundant HR policies and introduce innovative HR policies in the organization.

MAIN FOCUS OF CHAPTER

Innovation has been less studied as a function in HRM (Mishra et al., 2014; Darwish et al., 2019). However, its essence in organizational effectiveness is significant (Agarwala, 2003; Rao, 1990; Ulrich, 1997; Yeung & Berman, 1997). Thus, the basic objective to work on this chapter is to explain the innovations occurring in the field of HRM area to advance the work systems and encourage employees to engage in organizational activities with enthusiasm and motivation which can result in attaining sustainable competitive advantage for organizations. The chapter will discuss the innovative HR policies initiated in different sectors of Indian economy.

METHODOLOGY

Since the research in the field of innovation in HRM practices is not novel, thus, there is a vast amount of literature available. More than 100 research papers which included studies on innovation in HRM practices were pulled from various sources such as EBSCO, JSTOR, PROQUEST, Google scholar, academic info, BASE, Eric,

Citeulike, archival research catalogue, Infotopia, Refseek, the virtual LRC, Infomine, Microsoft academic search, and iSeek. The focus while doing the literature review on innovation in HRM practices was on the recent papers so that the current state of the field could be identified. While doing the literature review on innovation in HRM practices, though there were umpteen literatures available, only classical and relevant papers have been used. The keywords chosen to search the articles/ papers were innovative HR policies, HRM and innovation, and Innovation in HRM. Besides articles and papers, few surveys and newspaper articles were also referred.

INNOVATIONS IN HRM IN INDIAN ORGANIZATIONS

HRM in Indian organizations has attained a significant position from changing its position as salary distribution department to a fully fledged working department taking care of recruitment, selection, training, development, performance evaluation, promotion, salary, and other significant functions. With the influx of IT revolution in India and foreign investments in different sectors of Indian economy HRM has also improved itself from a function to a vast and diversified area to be researched and enhanced. HRM innovations are the result of implementing new and advanced HR policies in place of redundant ones. Many sectors of Indian economy has initiated these innovative HR policies including IT, Banking, Healthcare, Hospitality, Consultancy, Tourism, Food and Entertainment, and many more are in the line. To present some of the examples of Innovative HRM practices in Indian companies a table has been given below mentioning the Innovative HR policies and their outcome on company's performance. It will help in understanding the impact and essence of innovation in the field of HRM for Indian organizations.

FINDINGS OF THE STUDY

Above presented table of HR innovations in different sectors of economy states that, by implementing innovative HR practices, companies will be able to sustain change and uncertain business environment. Previous research work also stated that, innovative HR practices always affect positively towards organizational commitment of employees (Paul & Anantharaman, 2004). Practices like innovative and open work environment, facilities for career enhancement, growth oriented learning and training programs followed by an effective appraisal system are relevant for meeting the expectations of employees. Companies gain strategic advantage over their competitors by implementing innovative HR practices in the areas of recruitment and selection, training and development, learning and other facilities (Srivastava &

Table 1. Innovative HRM practices or policies in Indian organizations.

No.	Innovation	Company/Industry	Description	Outcome
1	GOLD (Godrej Organization for Learning and Development)	Godrej Industries Ltd. (Real estate, consumer products, appliances, furniture, etc).	Company initiated web-based learning in collaboration with UK – based NetG to distribute e- learning modules among the workforce. The company gives equal importance to soft skill training.	This learning creates a leadership pipeline
2	Recruitment of HR Professionals, Free flow of internal communication,	Maruti Udyog Ltd (Automobiles)	Company innovate its HRM functions to improve internal efficiency of the firm.	Company was able to reduce production costs and increase efficiency of the employees with smooth flow of communication and enhanced HR practices.
3	BPR (Business Process Re-engineering), Flat structures, Outsourcing workforce	Mahindra & Mahindra (Automobile)	Company initiated these schemes to reinvent business processes, encourage teamwork, knowledge sharing and learning from each other, and attain skilled workforce for advanced and non-core activities.	Company was able to attain steady and increased profits, and gradually manufacturing productivity increased.
4	Infosys Toastmasters Club, INSTEP (Leadership internship)	Infosys (IT)	To provide support to the employees. To develop leadership skills among employees	Company attained skilled and visionary workforce to handle uncertainties of global markets.
5	ESOP (Employee stock option programme)	Wipro (Consumer goods, IT)	To share the organizational membership within employees	Employees feel like members of family and dedicate themselves for accomplishing organizational goals.
6	The PEP (The Performance Ethic Program), Performance Management System	TISCO	It encourages young professionals to grow up the corporate ladder rapidly instead of following traditional promotion ways.	Company succeeded in attaining employees' confidence and trust in them.
7	Cross border learning programs	Ranbaxy (Pharmaceuticals)	It encouraged knowledge exchange facility for its employees so that they can learn about the advanced techniques from them and gain the knowledge about global markets.	Company was able to improvise its R&D facilities and invented new drugs for various significant diseases.
8	Redeployment and retraining of employees, revamped performance appraisal system	BPCL (Bharat Petroleum Corporation Ltd.1)/ Petroleum	Company regarded HRM as an important support service in employee engagement	Company was able to retain its customers and maintain profitability.
9	Introduced VRS (Voluntary Retirement Scheme)	SBI (State Bank of India)/ Bank	Company segmented HR by empowering employees to fulfil their duties and introduced VRS to manage strategic changes like automation.	Company reduced its workforce at rationalized costs and also compete with foreign banks.
10	CLAP (CLARIANT Participation to improve Profitability through Performance of People)	Clariant Ltd./ Chemicals	Company introduced CLAP program to change the mindset of employees through communication and goal settings	Company attained a transition from Sandoz to Clariant.
11	Innovative recruitment, compensation, and training procedures	Arvind Mills/Textiles	Company initiated innovative ways of recruiting, training and compensating its employees along with bridging the gap between different levels of management	Company succeeded in adapting change of fashion industry and increased demand of the products at global level.
12	Turnaround HR strategies, 'HAI DUM', Work Life Balance	Motorola India	Company focussed on innovating HR practices in learning and development, rewards, performance appraisal, recruitment and facilitating employee oriented work culture.	A continuous process of learning-unlearning and relearning was successfully implemented in the company.
13	Redefined organizational structure and optimally utilized HR resources	Mehta Group/ Cement	Company redefined organizational structure and able to implement HR policies to leverage its HR resources in an effective manner.	Developed synergy in the form of structure, manpower, and resources along with curtailing the competition among different companies of same group.
14	Gyan Jyoti (E-learning), HELLO (Helping employees launch and learn in the organization), NEST (Nurturing engagement with satisfaction and trust)	Tata Steel	Company introduced manager assimilation program along with e-learning initiatives to improve employees' skills, initiated induction and orientation programs to introduce new employees about the organization and nurturing their talent and skills for their personal and professional growth.	Tata has become one of the favourite employer brand of the aspiring and existing employees

continued on following page

Table 1. Continued

No.	Innovation	Company/Industry	Description	Outcome
15	360 Degree Performance Appraisal, Talent Identification and Development strategy	Hindustan Unilever Ltd. (HUL)	Company initiated performance appraisal technique to develop the employees and engage them in their organizational duties.	Company attained reduced attrition levels and increase in productivity and quality service from employees.
16	Strategic alignment of HR policies with organizational goals and objectives	SAIL (Steel Authority of India)	Company invested in various HR practices to nurture and develop the employees and facilitated them with conducive workplace where employees' creativity and innovation enhanced	Company attained success in attaining support of labour associations, in creating a peaceful and harmonious organizational climate.
17	Green HRM (GHRM)	BHEL(Bharat Heavy Electricals Ltd)	Company invested heavy amounts to attain sustainable business outcomes.	Company attained employee support and commitment with specific reference to corporate social responsibility.
18	HRIS (Human Resource Information System)	NALCO (National Aluminium Company Ltd)	Company implemented HRIS to maintain record of its employees for supporting HRD activities like knowledge enhancement, skill development, and attitude building.	An enhanced module helps in sufficing HRD facilities to organizational employees leading to organizational growth and sustainability.
19	Gyan Drishti (Learning Management System)	Jindal Steel	Company initiates this practice to encourage employees towards learning and knowledge sharing through a self learning platform.	Employees engage more in their activities, do brain storming sessions to solve issues and bring creative ideas for organizational sustainability.
20	Humanistic Attitude	Aditya Birla Group	Company reaped back the benefits for the growth and welfare of underprivileged sections of society.	Gained more humanistic attitude among employees with a sense of fulfilling corporate social responsibility and maintaining environmental sustainability.

Bhatnagar, 2008). However, implementing innovative services require acceptance from organizational employees and a strong leadership otherwise it can be easily transformed into negative consequences.

FUTURE RESEARCH DIRECTIONS

With the uncertain business environment and constantly changing customer demands, it has become utterly important for companies to transform their strategies from old school of thoughts to radical ones in order to compete not only locally but at global platforms also. In Indian economy not only big companies but small and medium sized enterprises (SMEs) are also enhancing the GDP growth rate thus along with big corporate houses, they also need to unleash their capacity and increase the productivity. However, the rule of thumb applies to all equally which says that like big industrial organizations, SMEs also need to innovate their HR policies. Till now the research focused on identifying the impact of innovative HR practices on organizational performance in big business houses but future research work conduct the same research on SMEs. Also apart from quantitative, qualitative studies can also be carried out to identify the innovation implemented in HR practices and their impact on employees' organizational commitment or organizational sustainability. Moreover, research can also find answer to the biggest challenge faced by HR leaders that is How to engage and retain employees through innovative HR practices? Research can also be initiated to analyze the impact of Green HRM or E-HRM on employee commitment or organizational performance.

CONCLUSION

This study has taken into account the HR innovative practices implemented in Indian organizations and their impact on organizational growth and sustainability. From the review of literature, a table has been prepared depicting the innovative HRM practices implemented in different Indian organizations and its consequences (Som, 2006). The study has a major limitation that it is descriptive in nature and based on review of literature. However, in upcoming future a qualitative or quantitative study can be conducted to enhance the scope of research in the field of Innovative HRM practices.

ACKNOWLEDGMENT

I would like to thank the authors of the book and members of publishing house to provide me this opportunity to present my work. This research received no specific grant from any funding agency in the public, commercial, or not-for-profit sectors.

REFERENCES

Agarwala, T. (2003). Innovative human resource practices and organizational commitment: An empirical investigation. *International Journal of Human Resource Management*, *14*(2), 175–197. doi:10.1080/0958519021000029072

Amba-Rao, S. C. (1994). Human resource management practices in India: An exploratory study. *Indian Journal of Industrial Relations*, *30*(2), 190–202.

Antoncic, B. (2007). Intrapreneurship: A comparative structural equation modeling study. *Industrial Management & Data Systems*, *107*(3), 309–325. doi:10.1108/02635570710734244

Antoncic, B., & Hisrich, R. D. (2001). Intrapreneurship: Construct refinement and cross-cultural validation. *Journal of Business Venturing*, *16*(5), 495–527. doi:10.1016/S0883-9026(99)00054-3

Bamber, G. J., Bartram, T., & Stanton, P. (2017). HRM and workplace innovations: Formulating research questions. *Personnel Review, 46*(7), 1216–1227. doi:10.1108/PR-10-2017-0292

Benitez-Amado, J., Llorens-Montes, F. J., & Perez-Arostegui, M. N. (2010). Information technology-enabled intrapreneurship culture and firm performance. *Industrial Management & Data Systems, 110*(4), 550–566. doi:10.1108/02635571011039025

Boxall, P., & Purcell, J. (2011). *Strategy and Human Resource Management* (3rd ed.). Palgrave Macmillan.

Boxall, P., & Purcell, J. (2016). Strategic HRM and sustained competitive advantage. *Strategy and Human Resource Management*, 82-103.

Cavanagh, J., Bartram, T., Meacham, H., Bigby, C., Oakman, J., & Fossey, E. (2017). Supporting workers with disabilities: A scoping review of the role of human resource management in contemporary organisations. *Asia Pacific Journal of Human Resources, 55*(1), 6–43. doi:10.1111/1744-7941.12111

Chand, M. (2010). The impact of HRM practices on service quality, customer satisfaction and performance in the Indian hotel industry. *International Journal of Human Resource Management, 21*(4), 551–566. doi:10.1080/09585191003612059

Collins, B. (2020). "It's not talked about": The risk of failure in practice in sustainability experiments. *Environmental Innovation and Societal Transitions, 35*, 77–87. doi:10.1016/j.eist.2020.02.008

Cooke, F. L., & Saini, D. S. (2010). (How) does the HR strategy support an innovation-oriented business strategy? An investigation of institutional context and organizational practices in Indian firms. *Human Resource Management, 49*(3), 377–400. doi:10.1002/hrm.20356

Damanpour, F. (1991). Organizational innovation: A meta-analysis of effects of determinants and moderators. *Academy of Management Journal, 34*(3), 555–590.

Darwish, T. K., Wood, G., Singh, S., & Singh, R. (2019). Human Resource Management in India: Performance and Complementarity. *European Management Review, 17*(2), 373–389. doi:10.1111/emre.12367

Datta, D. K., Guthrie, J. P., & Wright, P. M. (2005). Human resource management and labor productivity: Does industry matter? *Academy of Management Journal, 48*(1), 135–145. doi:10.5465/amj.2005.15993158

Fan, P. (2011). Innovation capacity and economic development: China and India. *Economic Change and Restructuring, 44*(1-2), 49–73. doi:10.100710644-010-9088-2

Guest, D., & Conway, N. (2011). The impact of HR practices, HR effectiveness and a 'strong HR system'on organisational outcomes: A stakeholder perspective. *International Journal of Human Resource Management*, 22(8), 1686–1702. doi:1 0.1080/09585192.2011.565657

GUPTA, M. (2016). Intrapreneurship centric innovation: A step towards sustainable competitive advantage. *International Journal of Innovative Research and Development*, 5(2), 90–92.

Gupta, M. (2018). The innovation process from an idea to a final product: A review of the literature. *International Journal of Comparative Management*, 1(4), 400–421. doi:10.1504/IJCM.2018.096731

Kong, E., Chadee, D., & Raman, R. (2013). Managing Indian IT professionals for global competitiveness: The role of human resource practices in developing knowledge and learning capabilities for innovation. *Knowledge Management Research and Practice*, 11(4), 334–345. doi:10.1057/kmrp.2012.21

Kramar, R. (2014). Beyond strategic human resource management: Is sustainable human resource management the next approach? *International Journal of Human Resource Management*, 25(8), 1069–1089. doi:10.1080/09585192.2013.816863

Kulkarni, S. V. (2013). Innovation management-challenges and opportunities in the next decade. *Asia Pacific Journal of Management & Entrepreneurship Research*, 2(1), 225–235.

Lengnick-Hall, M. L., Lengnick-Hall, C. A., Andrade, L. S., & Drake, B. (2009). Strategic human resource management: The evolution of the field. *Human Resource Management Review*, 19(2), 64–85. doi:10.1016/j.hrmr.2009.01.002

Lenka, U., & Gupta, M. (2019). An empirical investigation of innovation process in Indian pharmaceutical companies. *European Journal of Innovation Management*, 23(3), 500–523. doi:10.1108/EJIM-03-2019-0069

Lenka, U., Gupta, M., & Sahoo, D. K. (2016). Research and development teams as a perennial source of competitive advantage in the innovation adoption process. *Global Business Review*, 17(3), 700–711. doi:10.1177/0972150916630841

Malik, A., Boyle, B., & Mitchell, R. (2017). Contextual ambidexterity and innovation in healthcare in India: The role of HRM. *Personnel Review*, 46(7), 1358–1380. doi:10.1108/PR-06-2017-0194

Mandip, G. (2012). Green hrm: People management commitment to environmental sustainability. *Research Journal of Recent Sciences*, 1, 244–252.

Miles, R. E., & Snow, C. C. (1984). Designing strategic human resources systems. *Organizational Dynamics, 13*(1), 36–52. doi:10.1016/0090-2616(84)90030-5

Mishra, R. K., Sarkar, S., & Kiranmai, J. (2014). Green HRM: Innovative approach in Indian public enterprises. *World Review of Science, Technology and Sustainable Development, 11*(1), 26–42. doi:10.1504/WRSTSD.2014.062374

Muster, V., & Schrader, U. (2011). Green work-life balance: A new perspective for green HRM. *German Journal of Human Resource Management, 25*(2), 140–156. doi:10.1177/239700221102500205

Naim, M. F. (2014). Leveraging social media for Generation Y retention. *European Journal of Business and Management, 6*(23), 173–179.

Naim, M. F., & Lenka, U. (2017). Linking knowledge sharing, competency development, and affective commitment: Evidence from Indian Gen Y employees. *Journal of Knowledge Management, 21*(4), 885–906. doi:10.1108/JKM-08-2016-0334

Parker, S. C. (2011). Intrapreneurship or entrepreneurship? *Journal of Business Venturing, 26*(1), 19–34. doi:10.1016/j.jbusvent.2009.07.003

Paul, A. K., & Anantharaman, R. N. (2004). Influence of HRM practices on organizational commitment: A study among software professionals in India. *Human Resource Development Quarterly, 15*(1), 77–88. doi:10.1002/hrdq.1088

Renwick, D., Redman, T., & Maguire, S. (2008). Green HRM: A review, process model, and research agenda. *University of Sheffield Management School Discussion Paper, 1*, 1-46.

Rogers, E. (1983). *Diffusion of Innovations* (3rd ed.). The Free Press.

Ruël, H., Bondarouk, T., & Looise, J. K. (2004). E-HRM: Innovation or irritation. An explorative empirical study in five large companies on web-based HRM. *Management Review, •••*, 364–380.

Shafaei, A., Nejati, M., & Yusoff, Y. M. (2020). Green human resource management. *International Journal of Manpower, 41*(7), 1041–1060. doi:10.1108/IJM-08-2019-0406

Shipton, H., Budhwar, P., Sparrow, P., & Brown, A. (2017a). Editorial overview: HRM and innovation—a multi-level perspective. *Human Resource Management Journal, 27*(2), 203–208. doi:10.1111/1748-8583.12138

Shipton, H., Sparrow, P., Budhwar, P., & Brown, A. (2017). HRM and innovation: Looking across levels. *Human Resource Management Journal, 27*(2), 246–263. doi:10.1111/1748-8583.12102

Som, A. (2006). Bracing for MNC competition through innovative HRM practices: The way ahead for Indian firms. *Thunderbird International Business Review, 48*(2), 207–237. doi:10.1002/tie.20093

Srinivasan, V., & Chandwani, R. (2014). HRM innovations in rapid growth contexts: The healthcare sector in India. *International Journal of Human Resource Management, 25*(10), 1505–1525. doi:10.1080/09585192.2013.870308

Srivastava, P., & Bhatnagar, J. (2008). Turnaround@ Motorola India: Mobile Devices Business through the HR Lever. *Vikalpa, 33*(3), 121–142. doi:10.1177/0256090920080309

Stanton, P., Gough, R., Ballardie, R., Bartram, T., Bamber, G. J., & Sohal, A. (2014). Implementing lean management/Six Sigma in hospitals: Beyond empowerment or work intensification? *International Journal of Human Resource Management, 25*(21), 2926–2940. doi:10.1080/09585192.2014.963138

Ulrich, D. (1997). Measuring human resources: an overview of practice and a prescription for results. *Human Resource Management, 36*(3), 303-320.

Ulrich, D. (2016). HR at a crossroads. *Asia Pacific Journal of Human Resources, 54*(2), 148–164. doi:10.1111/1744-7941.12104

Wolfe, R. A. (1995). Human resource management innovations: Determinants of their adoption and implementation. *Human Resource Management, 34*(2), 313–327. doi:10.1002/hrm.3930340208

Yeung, A. K., & Berman, B. (1997). Adding value through human resources: Reorienting human resource measurement to drive business performance. *Human Resource Management, 36*(3), 321-335.

Chapter 4
The Impact of Social Media on Recruitment in New Age Organizations

Richa Das
Atria Institute of Technology, India

ABSTRACT

The new age world, which has become part of our lives, is a world of rapid innovations and changing technologies. These changes bring new opportunities for organizations to exchange information, news, ideas, and work. Attracting and retaining best of employees has become the most crucial tactical problems for the people's department of companies all over the world. In the current scenario of high competition, the internet has substantially converted the features of recruitment and selection procedure of the businesses. The chapter presents an exploratory research on the impact of social media on recruitment in new age organizations.

INTRODUCTION

As managers are becoming more aware that running a successful organization is decided by having the best workforce at the right place for unbeaten implementation of their strategy (Stahl et al.,2012), a big shift can be observed in the prospect of the role and significance of recruitment. The recruitment function is moving from being an expanded back end role to a key value driver in the rivalry of best candiadte (Hunt, 2014). The recruitment process consists of three succeeding phases, beginning with searching for the right candidate, continuing by addressing the right people, and ending with the selection of right employees.

DOI: 10.4018/978-1-7998-4180-7.ch004

The application of social media for the purpose of recruitment, is often referred as 'social media recruitment'. The process comprises distinct policies and allows a number of benefits and advantages. As the number of social media users are increasing every passing day, social media as one of the channels for recruitment is acquiring an impulse and thrust among the HR professions (Dutta, 2014; Singh & Sharma, 2014). This new trend is a result of organizations' ability of spotting the capabilities of these channels to engage candidates who are in all phases of job search. The application of social media as a support in the recruitment and selection process can provide an extensive range of advantages for the organizations. By using social media for recruitment, an access is sanctioned to an extensive scale of job seekers who are effortlessly reachable at any given period of time. Organizations realize that social media recruitment is helpful because it reduces the time to fill the vacancy, also it raises the standard of hire (SHRM, 2016; Jobvite, 2014). Recruiters are using platforms like LinkedIn, Twitter and Facebook to publish job vacancy advertisements, attract and hire candidates, and pre-screen these candidates (El Ouirdi et. al, 2016). Social media is likely to provide more speed, efficiency and a better capability to focus on and attract suitable applicants for the recruitment and selection process. Employers are able to collect a great deal of private information about potential candidates "as a source of candidates data in an attempt to enhance decisions related to recruitment and selection" (Gueutal, Kluemper & Rosen 2009).

A CareerBuilder study conducted on more than 2,500 employers, published in year 2014, show around 35% of respondents make use of social media to improve branding of their organizations. Out of those respondents, 21% are employing social media to recruit and research about the potential candidates, and while the other 18% are utilizing it to build up their employment brands. 93 percent of the employers apply social media to assist in their recruitment efforts (Jobvite, 2014).

Jindal and Shaikh (2014) established that 50 percent of the companies utilize social media in paid-for job advertising across various platforms and thirty seven percent broadcast job openings via tweets or alerts or avail free job advertising space across selected social media platforms (e.g. Twitter, Instagram and Facebook). In addition, thirty percent of employers expand their database of followers by reporting or posting methodical updates. 18% of the recruiter employ the social media not only to advertise vacancies but also to receive resume and application forms on the behalf of the organization. Only 7 percent of employers use social media to screen the relevancy of potential candidates on their social media pages (Jindal and Shaikh, 2014).

BRIEF OVERVIEW OF RECRUITMENT AND SELECTION THROUGH SOCIAL MEDIA

Attracting and retaining right candidates has become one of the most crucial and tactical issues of human resource management for most of the organizations all over the world (Schlechter, Hung & Bussin, 2014; Singh & Finn, 2003). Recruitment and selection are most important aspect of talent management which cannot be ignored at any given period of organization's life cycle.

Molnar, (2011) describes recruitment as "the process of finding, selecting, attracting and hiring qualified workforce to be employed inside an organization and contribute to the accomplishment of its goals and objectives". Philips describes (2012) "selection is the process of screening applicants to decide which candidates meet the job demands in terms of knowledge, skills, abilities and responsibilities". The recruitment of most worthy employee is connected to the fulfilment of activities that are financially challenging for the organization. Recruitment of the wrong candidate can be costly and can bring loss to the organization. For a low-level position, a wrong recruitment may cost a company double the employee's annual salary, increasing to almost six times the annual salary at higher levels (Armstrong, 2006; Houran, 2017). Consequently, recruiters all over world are now taking an extra effort to cope matters concerned with attracting the best candidates, recruitment and selection of employees in their organisation (Holland, Sheehan & Pyman, 2007).

The idea of a 'war for talent' was already been noted in the year 1998 (Chambers, Foulon, Handfield-Jones, Hanking & Michaels, 1998). In the present- day corporate world, organizations are finding it very difficult to discover the right employees for their organizations (Schlechter et al., 2014). The phrase 'competency deficit' is often applied to narrate what can be one of the grounds why organizations are facing problems in attracting brilliant employees for themselves (Herman, Olivio & Gioia, 2003). Michaels et al. (2001) explains that the components pushing the growing competitiveness for right candidates comprises the shift from the industrial age to information age. The change has developed the necessity for a completely separate or dissimilar expertise and an ever-growing pressure for high-level managerial technique and expertise, additionally an increasing inclination among workers to find new jobs more often than was the case in the past.

With an acknowledgement of the significance of recruitment chiefly the significance of sourcing, the finest technique for hiring stays arguable for the organizations all over the world (Houran, 2017). Determined by the type of candidates who require to be recognized and drawn towards the organization, there are several methods to find right candidates, all of which has its own strengths and weaknesses (Hunt, 2014). Nevertheless, recognizing and attracting the best applicants is considered as a tough task (Sinha & Thaly, 2013). Less than a decade ago, sourcing of candidate was mainly concentrated on documented research exercises, such as job advertisements on various medium, employers' websites and job boards, with almost no emphasis on social media (Breaugh, 2008).

Kaplan & Haenlein (2010) describe social media sites as, "applications that allow users to connect, by creating personal information profiles, inviting friends and colleagues to have access to those profiles, and sending e-mails and instant messages between each other". Madia (2011) explains social media sites as a "web-based services that allow individuals to construct a public or semi-public profile within a bounded system, articulate a list of other users with whom they share a connection, and view their list of connections".

Lehtimäki et al., (2009) splitted social media into five different groups, namely: blogs and podcasts, social networks, communities, content aggregators, and virtual worlds. Social networking sites have linked people all over the world and mostly people are operating these sites to look for best available vacancies in the market, besides the additional purposes of entertainment and associating with others (Madia 2011).

The use of technology in companies has influenced every dimensions of employment process, with technology supplying remarkable and powerful ways to strengthen company recruitment operations (Gregory et al., 2013). Along with various processes, social media can be very useful tool in people management, for hiring function as well (Tufts, Jacobson, & Stevens, 2014; Wolf, Sims, & Yang, 2014). Social Media Recruitment is becoming an important source for recruitment and a buzzword for the business world. The employment of technology in hiring function is becoming an essential requirement for recruiters desiring to have a competitive edge over others in the business world, and to attract rare critical talent (Deloitte Consulting LLP, 2014). Recruiters utilize platforms such as Twitter, LinkedIn and Facebook from posting job vacancies, to attracting and recruiting candidates, and to pre-screen the candidates (El Ouirdi et. al, 2016).

As the count of end users of social media is increasing, the utilization of social media by organizations for recruitment and selection is becoming more popular (Dutta, 2014; Singh & Sharma, 2014). This inclination towards social media is an outcome of organizations' ability of identification of the vast possibilities of above-mentioned new channels to engage not only active potential job candidates but also passive and semi-passive candidates. Social networking sites such as LinkedIn, Twitter and Facebook permit recruiters to post job advertisements to attract a whole new scale of prospective candidates to effortlessly access and register for such possible vacancies, therefore equipping employers to search for and screen possible job applicants – even those who do not definitely apply for the job vacancies (Sinha & Thaly, 2013).

By exploiting social media networks for recruitment, access is permitted to a wider set of candidates who are effortlessly accessible at any given period of time.

For example, now LinkedIn itself has more than three million active job postings (Chaudhary, 2017). Furthermore, using social media networks makes the approach a feasible option for recruiters at a steadily lower cost (Broughton, Foley, Ledermaier & Cox, 2013; Brown & Vaughn, 2011; Khullar et al., 2017). It is assumed that by implementing social media recruitment, the job vacancies can be filled much rapidly and accordingly conserve a lot of time which can be further utilized in looking for qualified candidates through the usage of traditional means of job advertising and job posting (Sweeny & Craig, 2011). Sweeny & Craig (2011) argued that social media sites can furthermore assist organizations to grow their brand visibility on internet which can set up an outstanding image and identification for these organizations. Social media promises higher speed, effectiveness and the magnitude to target and attract specific, especially most suitable applicants in the procedure of recruitment. It can supply with the most competent source of information on possible job seekers, particularly since few of the data available, at professional and personal level, may not be created with the motive and determination of recruitment, and consequently may produce informal additional information about the candidate.

The University of Applied Sciences in Wiesbaden performed a research among 200 German recruiters to find the degree of role of web 2.0 in their recruitment process (Jäger & Meser, 2007). They concluded that thirty seven percent of the respondents suppose that the social networking sites are crucial for the people management of an organization and 85% attribute its importance for the time ahead. In the year 2008, a study conducted for the US Society for Human Resource Management (SHRM) established that the total number of organizations that accounted employing social networking sites as a recruitment tool had increased significantly from 21% in 2006 to 44% in 2008. Almost thirty-four per cent of these organizations are utilizing the social networking sites as a promotion tool to hire or to get in touch with applicants and other thirteen per cent are utilizing them as a screening device for the organization (Davison, Maraist and Bing, 2011). In the USA, another study proves that ninety per cent of job aspirants think that organizations inspect their social media accounts prior to calling for a job interview or any vacancy (Simply Hired, 2012). A European study data exhibits that all the applicants under the age of 25 feels that the contact with employers should take place more and more online. The same study proposes that in United Kingdom, Facebook is presently most favored over LinkedIn (64% compared to 52%) by new age candidates (Potential park (undated), cited in Clements, 2012).

Many employers are of the opinion that employing the use of social media sites for online background checks is a more sustainable method for taking recruitment and selection decisions to receive a fast character outline of the candidate (Clark & Roberts, 2010). Profiles on social networking sites permit head hunters to acquire information about candidates' education and professional career extensively. Additionally, employers have more probability to establish communication with possible candidates and speak with them in an unrevealing fashion. Most of the candidates use social media sites as a root of information about companies. Organizations inversely utilize social media as a source of detailed information about candiadates. Aforementioned practice has become a frequent strategy for a recruitment procedure (Jäger & Porr, 2008). Above 95% of employers who operate social media for their hiring procedure specified that they make use of LinkedIn (Bullhorn, 2014), contrast to sixty six percent employing Facebook and other fifty two percent get involved with applicants on Twitter (Jobvite, 2014). The indicated inclination to use social media is also substantiated by Zide et al. (2014), who established that every respondent in their research use LinkedIn for hiring process of their organization. Moreover, head hunters who utilize LinkedIn more frequently for the purpose of sourcing for their organization have noticed more success and hence employ it more often (Caers & Castelyns, 2011; Ollington, Gibb, Harcourt & Doherty, 2013). As a matter of fact, the staffing and recruitment function of HR is the most utilized function on LinkedIn (Darrow, 2017). Houran (2017) established that LinkedIn is immensely utilized tool for the recruitment of applicants or job seekers for major management positions at senior level (87%) and middle management levels (80%), while it is very seldom applied for entry level positions (8%).

Above discussed research establishes the significance of social media in the hiring process for new age organizations across the world.

1. **Impact of Social Media Sites on Various Functions of Recruitment and Selection**: In this section, author is going to discuss some major impact of social media in the process of recruitment and selection. Following are the few domains where social media impacts most in recruitment:

    a. **Impact of Social Media on Hiring Cost**: Sourcing for applicants or passive job seekers via social media platforms is more cost–effective as compared to any of the traditional methods of recruitment. Previous data also displays that the utilization of different form of social media recruitment methods bring down the recruitment costs by around 87% as compared to usual traditional recruitment tools like advertising in newspapers and radio (Lee, 2005; Cober et al., 2000). The Internet has generated possibilities for online recruitment practices to become important for coming several years. Most of the organizations are harnessing internet and technology by posting several job openings on their own websites and different job portals. Consequently, advertising for external positions has become more economical, faster and authorizes companies to be available to an expansive range of people (Anderson, 2003; Brady et al., 2003; Hull, 2011).

b. **Impact of Social Media on Hiring Time**: The assistance of social media recruitment facilities decreases the time-to-hire by allowing recruiters the chance to attach the jobs vacancies online with a click and by permitting job seekers to answer punctually by finishing forms posted online or by merely attaching their CVs or resume to the given email ids (Barber, 2006). Employers trust that using social media sites for online background scrutiny is a suitable exercise for taking recruitment decisions to acquire a speedy personality outline of the candidate (Clark & Roberts, 2010). Pin et al. (2001) argued that, as reported by a study executed among 500 American companies, the dominant edge of social media recruitment is reduction in time spent to hire the right candidate. 86% of the companies accepted the truth that social media has helped in reduction of hiring time of the organization.

c. **Impact of Social Media on Talent Pool**: As per the research conducted and published by Barber (2006), online advertising has a broader range of reach- locally, nationally and internationally- for increasing the talent pool and for achieving the heterogeneity of the applicants in the organization. As a result, employers have a greater prospect to discover the best candidates for vacancies in their organization. Social media also allow access to passive candidates who are not vigorously searching for a job but may have extremely accomplished talent and skilled. According to Eisele (2006), 57% of the German companies understand the significannce of social media recruitment in expanding the number of applicants for a job vacancy which in turn open up a chance of more diversified candidates in the organization.

d. **Impact of Social Media on Productivity of Employees**: Research indicates that organizations which use social media for recruitment have superior, more effective and inventive personnel than organizations that avail traditional recruiting plans. This may be because people who regularly spends time on social networking sites are possibly more original, experimental and tech-savvy (Massplanner, 2015).

e. **Impact of Social Media on Competitive Advantage:** The internet and social media sites plays very important roles in business roles and functioning nowadays. Hence, utilizing social media for recruitment of applicants imparts an organization with required competitive edge over other organizations that may not be using social networking platforms for recruitment. Social media recruitment not only represents a business house as tech-savvy but also open to emerging shift all over the world (Massplanner, 2015). Hatter (n.d.) motivates organizations to produce themselves as a winner of the competition by publicizing themselves in social media platforms, motivating the present employees to connect so as to boost the business and to associate other professional establishments to expand the organization presentation and for recruitment of high-quality candidates.

2. **Limitations of social media recruitment**: Even though social media recruitment has become very famous among the organizations because of the benefits it brings along with itself, it also leads the way for some problems as well. Following are few of the limitations of social media recruitment:

    a. **Precision of Information**: Screening of candidates utilizing social media is normally not a formalized feature of the hiring process in the organization, which proves that it may not be possible to substantiate that the information secured is correct and precise or not (Davison, Maraist and Bing, 2011; Nigel Wright Recruitment, 2011). The kind of information the end user selects to share on social media sites can be modified or twisted due to worries for social likeness and which may in turn be firmly depending on the recognized audience.

    b. **Security and Ethical Issues:** The details obtainable through social media sites leads the way to a sequence of distinctive legal disputes. Social networking sites simply permit the possibilities for individual biases that can influence the screening and hiring decisions. Employers all over are not currently supposed to reveal any information regarding what details from social networking sites were utilized in taking screening decisions, which can authorize managers to differentiate against applicants (Brown and Vaughn, 2011).

    c. **Contacting the Right Candidates:** Starting off direct communication with the possible candidates on social networking sites can be troublesome. Almost all social networking sites, including Twitter and LinkedIn, doesn't permit to send a personal message to its users who are not connected with each other at that time. According to Privacy Rights Clearinghouse, many job seekers are scared that social media networks may disclose avoidable information about them to prospective recruiters and finish up stiffening their own privacy settings, not letting recruiters with any chance of beginning any sort of communication with the right candidate.

METHODOLOGY

An exploratory research was done with the aim of understanding the current situation in social media recruitment. HR recruiters working in IT industry and consultancy sector in Bangalore, India were selected as study context. Self-administered questionnaire and semi structured in-depth interview method were used to gather data from the respondents. Some questions of the questionnaire were adopted from various studies in similar topics.

Respondents Profiles

One hundred and fifteen HR professional from IT and consultancy sector based in Bangalore, India were selected for the study. The age ranged from 28 years to 45 years. The experienced ranged between 3 years to 20 years

Data Collection Process

Data was collected using self-administered questionnaire and semi structured in-depth interview. Random multistage cluster sampling was done to collect data. Each interview was conducted separately for the comfort of respondent and lasted on an average of fifteen minutes.

The objective of the research was to understand the impact of social media on recruitment process in new age organizations and to explore dimension like cost per hire, reduction in hiring time, impact on productivity and impact on performance of employees through social media recruitment.

FINDINGS AND DISCUSSION

The social media usage for recruitment has become a common practice among the recruiters in the new age organizations, it is very important to understand the impact of social media recruitment and reasons for the rise in number of social media recruitment.

Fig 1 shows the position of various organizations on using social media recruitment. All the organization under study has some social media presence and use social media recruitment. The extensive use of social media recruitment is done by 70% of the organization, the other 30% selectively use social media for recruitment.

While the most favored method of posting job advertisement still remains company websites (37%), of late job portals (22%), social media sites (22%) and employee referrals (17%) has gained popularity among the HR managers. The least favored method of posting advertisement has changed to recruitment consultancy (8%) (Fig 2).

According to Fig 3, LinkedIn (30%) has the highest social media presence for new age organizations, followed by Facebook (29%) and Twitter (24%). Instagram (15%) has a considerable presence. Pinterest (2%) has least number of organizational profiles.

Figure 1. Organizations' position on use of social media

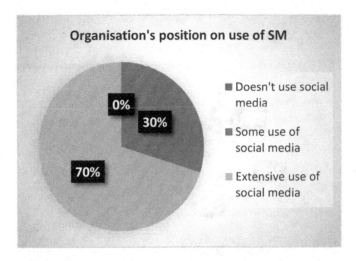

While most of the HR Managers think social media recruitment is cost effective (22%), they also believe SMR targets niche candidates (21%) and it accesses wider range of candidates (19%) in comparison to the other for of recruitment (Fig 4). The ease of use of social media (8%) remains the minimal reason for the managers, while the time saving and competitive advantage has been given equal weightage (15%) for the use of social media recruitment.

Figure 2. Favored method of vacancy advertisement

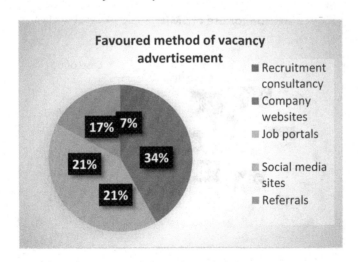

Figure 3. Organization Profile on social Media

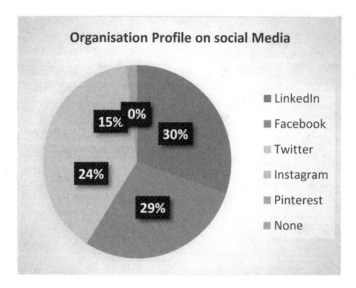

Most of the respondents use social media for job vacancy advertisement (49%), also 33% of HR Managers agree to use it to contact pool of potential candidates. Social media plays a big role in background check of candidates (12%) (Fig 5).

From Fig 6 it can be concluded that social media is maximum used to post jobs (49%) followed by employer branding (41%). The number of respondents using

Figure 4. Reason for use of social media recruitment

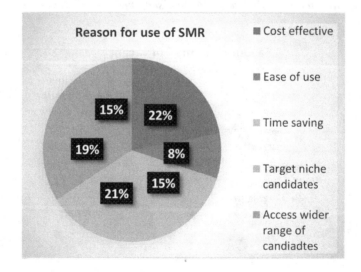

Figure 5. Recruitment stage in which social media is used

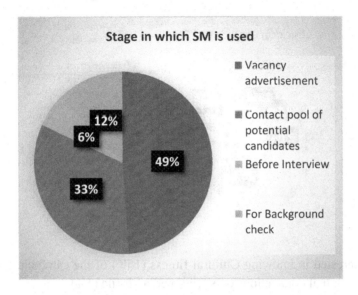

social media for employee referrals (8%) is very less whereas very nominal number of respondents use social media to vet candidates.

Basically, majority of HR managers look into the professional experience of candidate (37%) on social media, length of professional tenure in an organization (26%) and industry related posts also remain important variables (Fig 7). HR managers

Figure 6. Use of social media in recruitment process

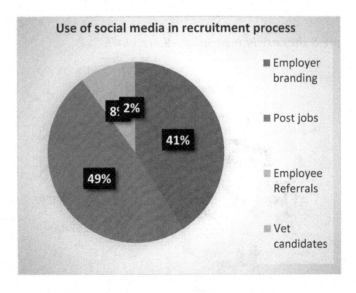

Figure 7. What is looked in candidates' social media

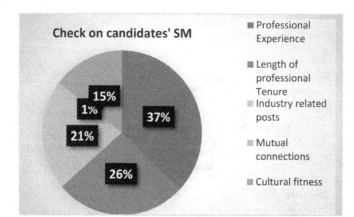

are also interested in knowing Cultural fitness (15%) of the candidates, very few look for the mutual connections on social media profile (1%).

Maximum number of times LinkedIn (67%) is used to conduct research about the candidates followed by Twitter (20%) and Facebook (15%) (Fig 8).

From Fig 9, it can be inferred that badmouthed previous employer (28%), inappropriate information/ photos and discriminatory comments (26%) on social media remain major reason for rejection of candidates. Poor communication skills on social media profile also plays very important role in rejection of candidates.

Fig 10 shows company monthly expenditure for fresh recruitment in most of the companies is highest for job portals followed by social media sites and lowest for recruitment consultancy.

Figure 8. Figure 8 sites used for social media research

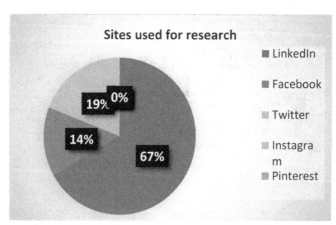

Figure 9. Reasons for rejection of candidates through social media

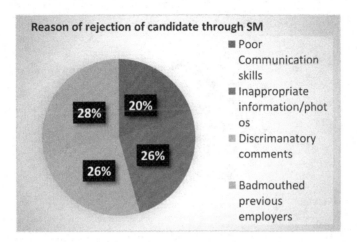

While 82% companies have a formal policy for social media recruitment (Fig 11), only 28% of respondents think they have expertise in social media recruitment (Fig 12). Followed by 60% of respondents feel they are proficient in social media recruitment and 12% consider themselves as novice in this skill. Possible discrimination based on personal choice has been cited as the biggest disadvantage (35%) of social media recruitment. This is followed by increased workload for HR mangers (25%) whereas candidates' privacy (20%) and disadvantage to candidates' who doesn't have access to internet (20%) also seems to be major concern for the respondents (Fig 13).

The quantitative survey gave the boarder picture of social media recruitment by the organizations, the qualitative survey of respondent description has led to

Figure 10. Monthly Expenditure for recruitment

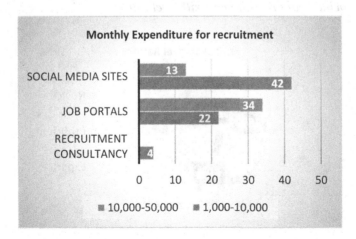

Figure 11. Formal policy for social media recruitment

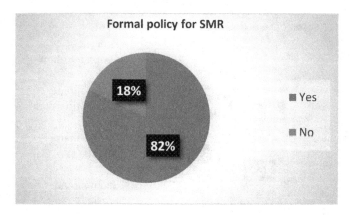

determination of different elements of success of social media recruitment. A comprehensive study of each subsequent topic is described with examples of quotes from the respondents in findings section. The rational understanding of the authors is also traced and presented.

Recruitment Cost

Recruitment cost is the total cost born by the organisation to bring in a new employee.
A respondent age 35, who works with a HR consultancy company said:

Social media is the best way to reduce the recruitment cost. Running job advertisements on social media is involves comparatively less expenditure than the traditional

Figure 12. Social media recruitment skill level rating

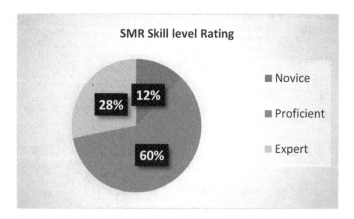

Figure 13. Limitations of social media recruitment

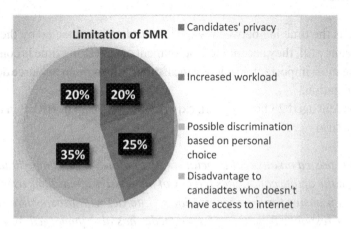

medium of recruitment. It is free to post job vacancy on Facebook and Twitter page of the company. Also, the reach of advertisement is much more on any social media. A simple ad on Facebook/Twitter can give twice more visibility than any other medium of recruitment. The cost per hire is much and engagement of a job posting is much higher on social media

The respondent clearly indicates that social media recruitment has bring down the recruitment budget and has increased the size of potential audience. Running an advertisement in magazines, billboards and TV is typically a more expensive method (Stammer, 2019).

Another respondent age 40, having an experience of more than 10 years in HR recruitment and working with IT company explained:

Social media recruitment is more cost effective. Organizations spent a big amount of money to post a job advertisement on job portals and then access resume of the potential candidates. Most of the job portals charge high for subscriptions. This cost is totally saved if the recruitment is being made by social media. In addition to this the return on hiring investment is significantly higher than any other medium.

From the above account it is clear, that other than traditional method of job posting among the new ones like job portals, social media recruitment is a cheaper medium of recruitment. According to Pandey, 2016, generally companies use a big amount of the organization's budget on recruitment. Integrating social media with recruitment will save a huge cost through mediums like job boards, job search or career portals.

Hiring Time

Hiring time is the time lag between when a candidate is contacted by the company for a job vacancy till, they accept the employment offer. Hiring time is considered to be one of the most important efficiency measures of the human resource department of the organization.

A respondent aged 28 having work experience of 2 years in HRBP in an finance consulatncy firm

When a job is posted on any of the social media platform it gets immediate response not only from the active candidates but few of the passive candidates as well. Social media makes it easier for the organization to interact with prospective candidates and get faster response from them. Social media has also immerged as a good way to conduct background check of the candidates, hence saving some backend time of the department.

The respondent clearly indicates that time involved in hiring process is reduced while using social media recruitment in comparison to any other traditional medium of recruitment. According to article published in Zippia (2018), Matson proposes that in comparison to traditional methods of recruitment, social media takes much less time.

Competitive Advantage

As explained by one of the respondents aged 37 responsible for recruitment and staffing in an IT company:

Social media has extended the playground for the companies. It is not only helpful in bringing best of the customers but also very helpful in attracting prospective employees through employee branding. It is also helpful in getting insights into potential candidates.

The respondent indicates that social media has become a great way to create brand image of the company and also attracting the potential candidates. According to Durfy (2018) if candidates are able to find company's social media profile either they either sought the company's profile or they have similar interest. If a candidate is ready to connect the organization based on the social media profile, it mostly means they are engaged in company's culture content making them better candidate in the talent pool.

Another respondent aged 45 working as Technical Recruiter for an IT company reports:

Social media has paved a way for the HR and candidates to find connections, common grounds which help them in developing mutual understanding which in turn helps in creating more trustful relationship resulting in increase in candidate-to-employee conversion rate

From the above account it can be concluded that Social media interaction between candidates and HR professional of the organization has laid a ground for successful interviews which has been argued in IES Blog.

Talent Pool

As discussed by one of the respondents aged 32, working in HRBP for a HR consultancy company:

Social media has proved itself as the best source of passive hiring. Since passive hiring requires more patience. It helps the HR manager to target such candidates and engage with them. The other important way through which social media has helped HR department is through getting more referrals, the present employees just have to share the job openings in your organization.

The respondent in above statement clearly points out that how contacting the right candidates has become easier with the presence of social media.

From the above discussion it can be concluded that social media recruitment has become very important tool for the recruitment and selection process of employees. It is evident that social media has helped the recruitment teams of organisation to improve overall perspective of recruitment. If the tool is used in a right manner it will help companies to realise its full potential.

CONCLUSION

From the above discussion it can be concluded that social media hiring is brings along great opportunities for recruitment. But it is not a solution all for the recruitment and selection problems faced by the organizations all over the world. Organizations can use the benefits of social media in terms of cost reduction, time consumption for hiring and position their business in a favorable spot; nevertheless, it is very important to pay attention to security and ethical issues which comes along with this answer.

REFERENCES

Anderson, N. (2003). Applicant and recruiter reactions to new technology in selection: A critical review and agenda for future research. *International Journal of Selection and Assessment, 11*(2/3), 121–136. doi:10.1111/1468-2389.00235

Armstrong, M. (2006). *A handbook of human resource management practice* (10th ed.). Kogan Page.

Barber, L. (2006). *e-recruitment Developments*. Retrieved 14.01.2020. from https://www.employment-studies.co.uk/pdflibrary/mp63.pdf

Blog, I. E. S. (2015). *The benefits of using social media as recruitment tool*. Available on https://www.innovativeemployeesolutions.com/blog/recruiters-and-staffing/the-benefits-of-using-social-media-as-a-recruitment-tool/#:~:text=Competitive%20edge,advantage%20over%20other%20recruiting%20firms

Brady, P. W., Thompson, L. F., Wuensch, K. L., & Grossnickle, W. F. (2003). "Internet recruiting", the effects of webpage design features. *Social Science Computer Review, 21*(3), 374–385. doi:10.1177/0894439303253987

Breaugh, J. A. (2008). Employee recruitment: Current knowledge and important areas for future research. *Human Resource Management Review, 18*(3), 103–118. doi:10.1016/j.hrmr.2008.07.003

Broughton, A., Foley, B., Ledermaier, S., & Cox, A. (2013). *The use of social media in the recruitment process*. Retrieved 20.12.2019, from https://www.acas.org.uk/media/pdf/0/b/The-use-of-social-media-in-the-recruitment-process.pdf

Brown, V., & Vaughn, E. (2011). The Writing on the (Facebook) Wall: The Use of Social Networking Sites in Hiring Decisions. *Journal of Business and Psychology, 26*(2), 219–225. doi:10.100710869-011-9221-x

Brown, V. R., & Vaughn, E. D. (2011). The writing on the (Facebook) wall: The use of social networking sites in hiring decisions. *Journal of Business and Psychology, 26*(2), 219–225. doi:10.100710869-011-9221-x

Bullhorn. (2014). *Global social recruiting activity report*. Retrieved 20.04.2019, from https://www.bullhorn.com/resources/2014-social-recruiting-activity-report/

Caers, R., & Castelyns, V. (2011). LinkedIn and Facebook in Belgium: The influences and biases of social network sites in recruitment and selection procedures. *Social Science Computer Review, 29*(4), 437–448. doi:10.1177/0894439310386567

Chambers, E. G., Foulon, M., Handfield-Jones, H., Hankin, S. M., & Michaels, E. G. I. I. I. (1998). The war for talent. *The McKinsey Quarterly*, (3), 44–57.

Chaudhary, M. (2017). *LinkedIn by the numbers: 2017 Statistics*. Retrieved 20.04. 2019, from https://www.linkedin.com/pulse/linkedin-numbers-2017-statistics-meenakshi-chaudhar

Clark, L. A., & Roberts, S. J. (2010). Employer's Use of Social Networking Sites: A Socially Irresponsible Practice. *Journal of Business Ethics*, *95*(4), 507–525. doi:10.100710551-010-0436-y

Cober, R. T., Brown, D. J., Blumental, A. J., & Levy, P. E. (2001). The quest for the qualified job surfer: It's time the public sector catches the wave. *Public Personnel Management*, *29*(4), 479–494. doi:10.1177/009102600002900406

Darrow, B. (2017). *LinkedIn claims half a billion users*. Retrieved 20.04.2019, from https://fortune.com/2017/04/24/linkedin-users/

Davison, H., Maraist, C., & Bing, M. N. (2011). Friend or Foe? The Promise and Pitfalls of Using Social Networking Sites for HR Decisions. *Journal of Business and Psychology*, *26*(2), 153–159. doi:10.100710869-011-9215-8

Deloitte Consulting, L. L. P. (2014). *Global human capital trends 2014: Engaging the 21st-century workforce*. Deloitte University Press.

Dutta, D. (2014). Tweet your tune – Social media, the new pied piper in talent acquisition. *The Journal of Decision Makers*, *39*(3), 93–104. doi:10.1177/0256090920140307

Eisele, S. (2006). *Online-Recruiting: Strategien, Instrumente, Perspektiven*. AV Akademikerverlag.

El Ouirdi, M., El Ouirdi, A., Segers, J., & Pais, I. (2016). Technology adoption in employee recruitment. *Computers in Human Behavior*, *57*, 240–249. doi:10.1016/j.chb.2015.12.043

Gregory, C. K., Meade, A. W., & Foster Thompson, L. (2013). Understanding internet recruitment via signalling theory and the elaboration likelihood model. *Computers in Human Behavior*, *29*(5), 1949–1959. doi:10.1016/j.chb.2013.04.013

Gueutal, H. G., Kluemper, D. H., & Rosen, P. A. (2009). Future employment selection methods: Evaluating social networking web sites. *Journal of Managerial Psychology*.

Hatter, K. (n.d.). *Competitive Advantage of Social Media.* Retrieved on 21.01.2020 from https://smallbusiness.chron.com/competitive-advantage-social-media-39239.html

Herman, R. E., Olivio, T. G., & Gioia, J. L. (2003). *Impending crisis: Too many jobs, too few people.* Oakhill Press.

Hired, S. (2012). *Today's Job Seeker Report. A survey of job seeker behaviors and motivations, 2012 Edition.* Available at http://success.simplyhired.com/rs/simplyhired/images/TodaysJobSeekerReport_2012_US.pdf

Holland, P., Sheehan, C., & Pyman, A. (2007). Attracting and retaining talent: Exploring human resources development trends in Australia. *Human Resource Development International, 10*(3), 247–262. doi:10.1080/13678860701515158

Houran, J. (2017). *New HR study: Candid recruitment experiences with LinkedIn.* Retrieved 20.04. 2019, from https://www.aethoscg.com/aethos_insights/new-hr-study-candid-recruitment-experiences-with-linkedin

Hull, J. (2011). 50% reduction on recruitment costs: how social media became my best friend. *HR Magazine.* Retrieved 21.01.2020, available at: www.hrmagazine.co.uk/hro/features/1019381/-reduction-recruitment-costs-social-media-friend

Hunt, S. T. (2014). *Common sense talent management: Using strategic human resources to improve company performance.* Wiley.

Jäger, W., & Meser, C. (2007). Personalmarketing 2.0. *Personalmagazin,* (12), 18–19.

Jäger, W., & Porr, D. (2008). Nutzenpotentiale des Web 2.0 im Personalmanagement. In *DGFP: Web 2.0 im Personalmanagement.* Retrieved 01.07.2019, from https://www.dgfp.de/media/content-downloads/546/web_2-0_pm.pdf

Jindal, P., & Shaikh, M. (2014). Social networking sites–Emerging as effective tools for attracting talent. *Gavesana Journal of Management, 6*(2), 48–55.

Jobvite. (2014). Social recruiting survey results. *Jobvite,* 1–17. Retrieved 20.04. 2019, from https://www.jobvite.com/wp-content/uploads/2014/10/Jobvite_SocialRecruiting_ Survey2014.pdf

Kaplan, A. M., & Haenlein, M. (2010). Users of the world, unite! The challenges and opportunities of Social Media. *Business Horizons, 53*(1), 59–68. doi:10.1016/j.bushor.2009.09.003

Khullar, A., Pandey, P., & Read, M. (2014). Effective use of social media recruiting. *International Journal of Management*, 4(4), 216–227. https://www. inderscienceonline.com/

Lauren, D. (2018). *6 compelling stats on social media recruitment.* Available at https://www.postbeyond.com/blog/6-social-media-recruiting-statistics/

Lee, I. (2005). The evolution of e-recruiting: A content analysis of Fortune career web sites. *Journal of Electronic Commerce in Organizations*, 3(3), 57–68. doi:10.4018/ jeco.2005070104

Lehtimäki, T., Salo, J., Hiltula, H., & Lankine, M. (2009). *Harnessing Web 2.0 for Business to Business Marketing- Literature Review and an Empirical Perspective from Finland.* Faculty of Economics and Business Administration, University of Oulu Working Papers, 29. Available at http://jultika.oulu.fi/files/isbn9789514291203.pdf

Madia, S. A. (2011). Best practices for using social media as a recruitment strategy. *Strategic HR Review*, 10(6), 19–24. doi:10.1108/14754391111172788

Marquis, M. (2018). *Advantages of using social media in your recruitment strategy.* Available at https://www.zippia.com/employer/8-advantages-using-social-media-recruitment-advertising-strategy/

Massplanner. (2015). *Benefits of Recruiting via Social Media.* Available at: http:// www.massplanner.com/benefits-recruiting-via-social-media/

Michaels, E., Handfield-Jones, H., & Axelrod, B. (2001). *War for talent.* Harvard Business School Publishing.

Molnar, W. (2011). *Human Resources Management, Recruitment Procedures* (3rd ed.). Create Space Inc.

Nigel Wright Recruitment. (2011). *The impact of social media on recruitment, Nigel Wright Recruitment, Report 2011.* Available at http://uk.nigelwright.com/ NigelWrightNews/2011-02-01/New-report-uncovers-true-impact-of-social-media-on-recruitment/

Nikolaou, I. (2014). Social networking web sites in job search and employee recruitment. *International Journal of Selection and Assessment*, 22(2), 179–189. doi:10.1111/ijsa.12067

Ollington, N., Gibb, J., Harcourt, M., & Doherty, R. (2013). Online social networks: An emergent recruiter tool for attracting and screening. *Personnel Review*, 42(3), 248–265. doi:10.1108/00483481311320390

Pande, S. (2016). *Social media can help mitigate risks and save costs in recruitment.* Available at https://www.beroeinc.com/article/socia-media-recruitment/

Phillips, J. (2012). *Accountability in Human Resource Management (Improving Human Performance)* (1st ed.). Routledge.

Pin, J. R., Laorden, M., & Sàez-Diez, I. (2001). *Internet Recruitment Power: Opportunities and Effectiveness.* Retrieved 01.09.2019, from https://www.iese.edu/research/pdfs/di-0439-e.pdf

Potentialpark. (2011). *Talent interaction: Does Facebook beat LinkedIn?* https://www.potentialpark.com/wp-content/uploads/2011/12/Potentialpark-Social-Media-Release-Europe.pdf

Schlechter, A., Hung, A., & Bussin, M. (2014). Understanding talent attraction: The influence of financial rewards elements on perceived job attractiveness. *SA Journal of Human Resource Management, 12*(1), 1–13. doi:10.4102ajhrm.v12i1.647

Singh, K., & Sharma, S. (2014). Effective use of social media for talent acquisition and recruitment. *International Journal of Intercultural Information Management, 4*(4), 228–237. doi:10.1504/IJIIM.2014.067932

Singh, P., & Finn, D. (2003). The effects of information technology on recruitment. *Journal of Labor Research, 24*(3), 395–408. doi:10.100712122-003-1003-4

Sinha, V., & Thaly, P. (2013). A review on changing trend of recruitment practice to enhance the quality of hiring in global organizations. *Management, 18*(2), 141–156. https://hrcak.srce.hr/index.php?show=clanak&id_clanak_jezik=166315

Stahl, G. K., Björkman, I., Farndale, E., Morris, S. S., Paauwe, J., & Stiles, P. (2012). *Global talent management: How leading multinationals build and sustain their talent pipeline (No. 2007/34/OB).* INSEAD.

Stammer. (2019). *Social media impact on recruitment.* Available at https://strammer.com/en/social-media-impact/

STATISTA. (2018). *Number of Social Network Users Worldwide From 2010 to 2021 (in billions).* Available at: https://www.statista.com/statistics/278414/number-of-worldwide-social-network-users/

Sweeny, S., & Craig, R. (2011). *Social Media for Business.* Maximum Press.

Tufts, S. H., Jacobson, W. S., & Stevens, M. S. (2014). Status update: social media and local government human resource practices. Academic Press.

Wolf, M. V., Sims, J., & Yang, H. (2014). Social media utilization in human resource management. In *Web based communities and social Media 2014 conference (WBC 2014); 8th multi conference on computer science and information systems*. Academic Press.

Zide, J., Elman, B., & Shahani-Dennig, C. (2014). LinkedIn and recruitment: How profiles differ across occupations. *Employee Relations*, *36*(5), 583–604. doi:10.1108/ER-07-2013-0086

APPENDIX

List of Questions

1. Which social media sites does your company have profile on?
    a. LinkedIn
    b. Facebook
    c. Twitter
    d. Instagram
    e. Pinterest
    f. None
2. What would be your favoured method of advertising a new role?
    a. Recruitment consultancy
    b. Company website
    c. Job portals
    d. Social media sites
    e. Referrals
3. Which of the following best describes your organisation's position on using social media for recruitment?
    a. My organisation doesn't use social media for recruitment
    b. We make some use of social media when recruiting staff
    c. We make extensive use of social media when recruiting staff
4. Why does your organisation use social media for recruitment?
    a. It is Cost effective
    b. Ease of use
    c. It saves time
    d. To target niche candidates
    e. To access wider range of candidates
    f. Competitive advantage
5. At what stage does your organisation use social media for recruitment
    a. To advertise job vacancies
    b. To make contact with pool of potential candidates for job vacancy
    c. Before interview
    d. To perform background check before offering the position
6. Why does your organization use social media for recruitment?
    a. Showcase employer brand
    b. Post jobs

      c.   Generate employee referrals

      d.   Vet candidates pre/ post interview

7.   What do you look for in a candidate on social media network?

      a.   Professional experience

      b.   Length of professional tenure

      c.   Industry related posts

      d.   Mutual connections

      e.   Cultural fit

8.   Which of the following sites would you use to research your interviewees prior to inviting a job interview in your organization?

      a.   LinkedIn

      b.   Facebook

      c.   Twitter

      d.   Instagram

      e.   Pinterest

      f.   None of the above

9.   What is the basis of rejecting job applications through social media check?

      a.   Poor communication skills

      b.   Inappropriate information/ photos

      c.   Discriminatory comments

      d.   Badmouthed previous employers

10.   What is your monthly expenditure in INR (1000-10,000)/ (10,000-50,000) for the following recruiting tools?

      a.   Recruitment consultancy

      b.   Job portals

      c.   Social media sites

11.   Does your organization have a formal policy covering the use of social media for recruitment?

      a.   Yes

      b.   No

12.   Rate your social recruiting skill level.

      a.   Novice

      b.   Proficient

      c.   Expert

13.   According to you what is limitation of social media recruitment

      a.   Candidates' privacy

      b.   Increase in workload

      c.   Possible discrimination based on personal choice

      d.   Disadvantage to candidates who don't have access to social media

Chapter 5
Problems and Prospects of Social Media Recruitment

Teena Saharan
Doon Business School, India

ABSTRACT

Over the last few years, the way of talent acquisition has evolved in different forms from attracting personal applications to getting connected with talented candidates through social networking sites. Recruitment through social networking platforms is putting a significant contribution in analyzing and hiring the right and best talent for an opening, and companies can't just ignore the potential and influence of these media platforms. These social platforms connect companies to potential hires and increase visibility by getting them connected to a huge audience. The future of recruitment lies in social media and companies cannot just ignore their presence due to prevailing challenges. It is important to find out viable solutions to the challenges organizations facing while using social media platforms in talent acquisition. The focus of this chapter is to capture strategies mitigating these challenges and suggest probable and profitable suggestions to companies for better utilization of social networking sites for effective recruitment.

RECRUITMENT AND DIFFERENT MODES OF RECRUITMENT

From the time that humans first began creating organizations, groups have been working on ways to be more effective at recruitment. In the 21st century, of course, most of the organization's efforts in recruiting went online. Obviously, this has not always been the case. In fact, the history of recruitment is almost as old as the history of civilization itself. For example, to complete the massive undertakings for the

DOI: 10.4018/978-1-7998-4180-7.ch005

Great Pyramids and Great Wall (Erin Engstrom, 2015), workers were handpicked in ancient Egypt and China.

The first instance of a resume can be found on rock and wooden tablets, dating back to ancient Rome, which had an engraving of what a person worked on, which was the first listing of the professional details of people. Another early evidence of recruitment could be seen in the history of imperial China. Imperial exams were a way of recruiting civil service candidates during the Han dynasty era around 1500 BC. These were one of the toughest assessments for centuries and often termed 'exams from hell' by the Britishers. The rise of recruiting did not really start to take shape until after World War II. To help those returning home from the war, new businesses sprouted that worked with job- seekers to find new opportunities. In time, organizations also contracted recruiting companies to help them find the perfect applicants for open positions. With the rise of the internet, these efforts have moved online, creating the job recruitment landscape seen every day (Emily Lennox, 2017).

With the increase in globalization, industrialization, consumerism, and economic growth, the need for the right talent increased many-fold. While these agencies were helping candidates by looking for the right kind of job for them, they realized it made more sense to hire candidates for companies—candidate, who had special skills and were required for a particular industry. It was at this time when recruitment agencies started to tie-up and collaborate with industries and help companies to hire the candidate for them specifically. They used to get job requirements or job descriptions, post them on media and job boards, interview candidates, and hire only the impressive ones. By late 1970s and early 1980s, database and online storage were introduced to the companies, giving them an immense supply of candidates who were interested in working with them. With the huge database and development in the telecom industry, suitable candidates could be found within a few minutes, making it easy for companies to invite them for the hiring process.

The first public job search engine went in operation in 1994. Monster Job board was created, a place where aspirants could browse the job database on internet. Monster Job disrupted the recruitment industry with its vision and ease of use. With the explosion of job boards and the movement of recruitment from print media to online channels, print ads for job requirements have almost become outdated.

The explosion of e-recruitment agencies gave recruiters new ways to reach a global market of candidates. With e-recruitment agencies helping globalize, a new requirement solution to automate the available database popped up, giving way to what we call the ATS (Applicant Tracking System). By the end of the 1990s, the Application Tracking System became a common industry term. The leading and foremost function and service of an ATS is to contribute and come up with a central location and also a database for a company's recruitment effort where the complete information from sourcing to hiring for the candidate was floated. Most of the e-recruitment agencies decided to upgrade themselves with the evolving requirements, converting them into a fully automated recruitment solution — ATS. With Web 2.0 and the growth of upgrading and enhancement of the social media and also the usage of mobile technology, the recruiter's ways to approach candidates changed. Now a Twitter/Facebook post with a URL link to the job description is shared with the candidates. Employers are wooing employees by video ads and slides. Many consider this recruitment broken, but I believe it has been always broken. Moving from pen-and-paper to agencies to ATS does not fix the recruitment. A social recruitment strategy does not mean posting a URL; rather it would mean an end-to-end recruitment cycle via social channels or the least by a cell phone. A fixed recruitment process would mean a process where a smart system screens candidate performance; an automated system evaluates their knowledge without any bias and an ATS that hires a candidate in a "true social sense" from media platform.

SOCIAL MEDIA

Social media is determined as the use of internet-based conversational media (applications that change the creation and transmission of content within the format of words, pictures, videos and audios) among communities of individuals who meet online to share information, knowledge and opinions (Safko & Brake, 2009). Four key motivations drive the utilization of social media: connect, create, consume and control (Hoffman & Fodor, 2010). Many well established social media platforms are available, for instance, Facebook, LinkedIn, Instagram, and Twitter etc. However, an outsized body of previous research indicates that among the varied social media platforms, Facebook, LinkedIn and Twitter are utilized in the sourcing process (Caers & Castelyns, 2011; Doherty, 2010; Dutta, 2014; Singh & Sharma, 2014). LinkedIn and Facebook might be classified as social networking platforms, in other words, platforms that allow users to share information about them, often through a web profile that they need created themselves (Safko & Brake, 2009). Twitter falls under the subcategory of micro-blogging tools, which permit users to share audio messages. Social networking and micro-blogging sites have shown tremendous growth over the past few years, with Facebook witnessing a mean of 1.32 billion daily active users in June 2017 (Facebook, 2017), a rise of 23 percent from the figure in 2016 (Zephoria, 2017), whereas LinkedIn had 467 million members in 2017 and Twitter had 317 million users (Chaudhary, 2017). Approximately 1 million professionals have published a post on LinkedIn and therefore the average user spends 17 minutes monthly on LinkedIn (Chaudhary, 2017).

Social Media Recruitment

Social media marketing is a buzz word and by 2020, it will be in the list of most demanding HR skills along with data analytics and predictive modeling. Social networking has dramatically changed the game of recruitment and recruiters are connecting using widely known sites such as LinkedIn, Facebook and Twitter. Using social media can help companies in finding out both active and passive candidates from the vast sea of hidden talent pool. Approximately 82 percent of the companies in USA alone and 78 percent of the companies worldwide are either recruiting through social media or thinking of using it to get best talent. Similarly, 79 percent of the job seekers use similar social platforms to search job options. 92 percent of the social media users prefer trusted recommendations over traditional marketing and promotion gimmicks of companies.

Social networking sites help companies to build brand, research about potential candidates, create awareness, list jobs, building direct connect with candidates, selecting better fit for jobs and getting references and tags on social posts. It's a win-win game for both- recruiters as well as candidates. Employees coming through social references have high convergence rate, stay long, and have higher job satisfaction in comparison to other recruitment sources. Social media saves a ton of company's money by providing them socially engaged employees which are more connected, inspired and optimistic. Businesses are able to find candidates on the basis of preferences, passion, networks or ideologies, the services that other platforms are unable to provide. Companies are communicating with shortlisted candidates using messenger, audio clips or chat box etc. It gives a personal touch to the process and candidates admire the strategy in the world of cold calls. It adds a special feel and catches the attention of even the passive candidates.

For a business to succeed, hiring best talent is of utmost importance which is a challenge for small entrepreneurs, not so famous enterprises and start-ups. Luckily, social media provides a viable platform to these companies to connect with potential candidates with matching interests. These platforms not only facilitate connection, it provides an opportunity to share corporate values, culture, working environment and much more. Almost 95 percent of the employers show social media presence and post information about their company. To sail from rudimentary to advanced recruitment tools, navigation is very important in the vast social networking space. Algorithm based technologies and artificial intelligence help in finding best talent as per requirement, analyze and rank them up on the basis of given job description. Analytics tools such as Talent Insight help companies indentify poaching targets from mix of passive candidates. These tools assist companies in estimating future skill demands and finding suitable candidates. These products and tools do much more than just locating the potential candidates along with their contact information such as email, phone number and location and integrate the data in firm's Applicant Tracking System (ATS).

Social media and online platforms like Facebook, Twitter, and Google AdWords offer new opportunities for researchers to recruit study participants for clinical research studies. The Pew Internet and American Life Project survey of 'Social Media Use' found that in 2016 Facebook was the most immensely used social network in United States. 76 percent Facebook users and 51 percent Instagram users (51%) visit these sites at least once a day. Facebook is also the most popular platform for recruitment. A search on PubMed in April 2017 using the keywords 'social media' and 'clinical trial' resulted in 267 studies mentioned social media for recruitment. Of those studies, 130 (48 percent) used Facebook for participant recruitment. According to a 2014 Pew Research Poll, Facebook is the top platform, regardless of race or ethnicity, with around seven-in-ten adult internet users (71 percent) reporting they browse the site; some of the lesser-used platforms vary in usage rates between different races and ethnicities (Ramo, 2017). Social media and recruitment research have stipulated that placing a billboard in popular media or on an organization's website features a limited chance of attracting the proper candidates (Phillips & Gully, 2012). This is often because mainly active and online candidates seem to use to those advertisements, leading to a little and limited candidate pool. Because the number of users on social media increases, the utilization of social media channels in recruiting is gaining momentum (Dutta, 2014; Singh & Sharma, 2014). This trend results from organizations' ability of recognizing the potential of those channels to draw in not only active prospective job candidates but also passive and semi-passive candidates. Social networking sites like LinkedIn, Facebook and Twitter allow recruiters to post job advertisements to lure a good spectrum of potential candidates to simply access and apply for such potential positions, thereby enabling recruiters to look for and screen potential job applicants – even those that don't necessarily apply (Sinha & Thaly, 2013).

It is evident that recruiters and organisations are realizing that more and better candidates are often discovered and approached quicker at a lower cost by utilizing social networks, compared to traditional recruitment methods (Armstrong, 2006; Singh & Sharma, 2014). Specifically, a robust association has been found between the utilization of LinkedIn and therefore the ability of identifying and attracting passive candidates (Nikolaou, 2014). By using social networks for recruiting, access is enabled to a good range of candidates who are easily accessible at any given time. For instance, LinkedIn now has 3 million active job listings (Chaudhary, 2017). Moreover, utilizing social networks makes this access possible at significantly lower cost (Broughton, Foley, Ledermaier & Cox, 2013; Brown & Vaughn, 2011; Khullar et al., 2014). It's especially the widely sought-after pool of very competent but passive candidates whom social media gives access to the recruiters (Doherty, 2010; Joos,

2008). That's why; it is not surprising that recruiters and organisations consider social media and networks as attractive recruitment tools which give recruiters a competitive edge up reaching their recruitment objectives (Singh & Sharma, 2014). It's evident that the utilization of social media platforms has become the norm for sourcing in recruitment and recruiters think that social media enables them to seek out better candidates. Indeed, 93 percent of recruiters use social media to support their recruiting efforts (Jobvite, 2014).

Social media is employed in recruitment for a variety of reasons. When inquiring into how recruiters use social media for advertising, Jindal and Shaikh (2014) found that 50 percent use social media in paid form for job advertisements and 37 percent advertise vacancies free of cost via tweets or alerts, or make use of free job advertising via targeted social media platforms such as Facebook. Furthermore, 30 percent of recruiters develop a database of followers and/or supporters by posting regular updates, and 18 percent use the social media platform's job search engines to advertise vacancies or to simply accept CVs and application forms. Surprisingly, only seven percent of recruiters use it to screen the suitability of potential recruits on their social networking pages.

Although Facebook is the biggest social media platform across the globe, however, it's not very mainstream or effective platform for recruitment (Bullhorn, 2014). Jobs posted on LinkedIn receive more views from potential candidates than those on Facebook and Twitter combined, and these posted jobs garner twice as many applications per general job advertisement. LinkedIn is the most preferred social network when it involves recruitment (Jobvite, 2014). Over 95 percent of recruiters who use social media in their recruitment process indicated that they use LinkedIn (Bullhorn, 2014), compared to 66 percent utilising Facebook and 52 percent engaging with candidates on Twitter (Jobvite, 2014). This trend is confirmed by Zide et al. (2014), who found that each one of the respondents in their research utilise LinkedIn in their recruitment process. Furthermore, recruiters who use LinkedIn more frequently within their sourcing have seen more success in the use of LinkedIn and thus use it more often (Caers & Castelyns, 2011; Ollington, Gibb, Harcourt & Doherty, 2013). Indeed, the staffing and recruitment industry is the one that is connected most on LinkedIn (Darrow, 2017). Houran (2017) found that LinkedIn is overwhelmingly utilized in the recruitment of candidates for key management positions at senior levels (87 percent) and middle management levels (80 percent), whereas it's very rarely used for entry positions (eight percent). These studies confirm the importance of LinkedIn within the recruitment process.

Apparently one among the most suitable use of LinkedIn among the social networking sites relates to its being seen by the general public as almost solely and exclusively for building professional relationships, which isn't the case with Facebook and Twitter, which are more general social media (Zide et al., 2014). Although all three of those social media platforms are getting used within the sourcing process, they have a tendency to be used differently. LinkedIn is usually used for posting advertisements, checking out candidates, and contacting and vetting candidates (Jobvite, 2014). On the opposite hand, Facebook and Twitter are used more to showcase the employer brand and to get referrals also on post advertisements (Jobvite, 2014). There is a marked preference among recruiters and human resource professionals for LinkedIn instead of Facebook for recruitment as they consider the previous to be simpler than the later (Nikolaou, 2014). Recruiters are of the opinion that LinkedIn gives the foremost insight into candidates' employment history, education, years of experience and how candidates portray themselves (Zide et al., 2014).

The Jobvite Annual Social Recruiting Survey 2014 is arguably one among the foremost comprehensive surveys of its kind. This survey was conducted online and was completed by 1855 recruiting firms and human resources professionals, spanning across several industries. The difference within the application of the social media platforms also yielded different results for recruiters. Overall, 79 percent of recruiters indicated that they placed a candidate through LinkedIn, 26 percent indicated that they did this through Facebook and only 14 percent indicated that they placed a candidate through Twitter (Jobvite, 2014). In spite of the very fact that Twitter is not the foremost popular social media platform in recruitment worldwide, an outsized body of research has shown that Twitter is employed extensively internationally by recruiters within the sourcing procedures (Caers & Castelyns, 2011; Dutta, 2014; Singh & Sharma, 2014). Although LinkedIn has one among the very best success rates of any website, it seems that it's still used but more conventional recruitment platforms like job boards, career portals and company websites, or a minimum of together with it (Allden & Harris, 2013; Tyagi, 2012). One of the reasons why many recruiters still prefer the more conventional sourcing tools could possibly be ascribed to recruiters' limited knowledge of the way to recruit effectively on social networking sites (Allden & Harris, 2013). Web-based job portals generate many applications, however, they don't necessarily reach to all the candidates, especially passive or semi-passive candidates (Sinha & Thaly, 2013).

Notwithstanding the substantial increment and increase within the utilization of social media for recruitment, one need to be careful to think that it is the panacea that resolves and settles all recruitment problems; it also has pitfalls (Doherty, 2010). For instance, with candidates uploading their own profiles it is likely to steer to profile inflation – which may be described as attempts to artificially enhance one's profile through little white lies or using deceptively positive terms to explain oneself, past accomplishments or current status (Houran, 2017). With the practice of candidate identification and screening through social media, legal implications are likely to arise thanks to the incorrect use of data (Melanthiou, Pavlou & Constantinou, 2015). Social media is extensively used for screening candidates; it's still unclear whether this screening influences a recruiter's decision to such an extent that they might reject an applicant (Melanthiou et al., 2015). Dutta (2014) argues that while social media offers various sourcing opportunities to recruiters, leading to a profound impact on the way that recruitment functions within the organizations, it should not be confused or mistaken for a full recruitment strategy. Rather, it forms merely a neighborhood of an organization's recruitment strategy. It might be argued that the recruitment process has been transformed from a mainly sequential process to a parallel process. Within the later process, social media tools cannot directly replace traditional sourcing tools, rather supplementing them the access to the highly sought-after passive candidate pool (Joos, 2008) and present them as active candidates (Doherty, 2010).

The massive widening of social media and internet capabilities has added numerous other sourcing possibilities and activities. Some of them are Internet Job Boards that allow organizations to upload their vacancies and candidates to upload their CVs, Internet Data Processing using Boolean searches and Web Crawlers/programmers that continuously search the online data for information about employees (Nikolaou, 2014; Parez, Silva, Harvey & Bosco, 2013; Sinha & Thaly, 2013). Flip searching is a process which identifies specific websites to look for passive and semi-passive candidates and social networking forleveraging connections on social media like LinkedIn, Facebook and Twitter. In this world of surplus Internet and social media where numerous options available, choosing which sourcing tool to use becomes an important decision within the recruitment process for any organization or recruiter (Galanaki, 2002; Sinha & Thaly, 2013). To draw-in high-caliber, passive and semi passive potential candidates, it becomes necessary to abandon the normal 'spray and pray' approach and to embrace the new sourcing tools offered by the web and social media (Dutta, 2014). Within the process, it is decisive to require cognizance of differences within the approach and philosophy between conventional and various social media recruitment tools (Dutta, 2014).

With so many benefits present, yet the biggest obstacle discouraging the use of this powerful medium is limited knowledge of social media recruitment. Simultaneously other inevitable issues challenge the social media recruitment of companies such as more traffic on social media (Facebook have 400 million users), low reach and engagement of company's social media page. Other than these, inauthentic job posts, costly social recruitment solutions (LinkedIn charges $195 for 30 days job post), disinterested employees in promoting organizational referral programs, legal complications related of social vetting and limited reach (unable to connect with economically weaker section) are other major pain areas for companies in social media recruitment.

TOOLS FOR SOCIAL MEDIA RECRUITMENT

This past year has witnessed an explosion of individuals on social networking sites round the globe. Groups and networks are created virtually for every profession, hobby and institution imaginable. At the present, this trend shows no sign of abating. Recent research from networking site Viadeo shows just how popular and well-known the web has become as a recruitment tool. One in five of the 600 employers it questioned admitted to finding information on potential employees online and 59 percent admitted that social media sources have influenced their decision-making. One-fourth of the recruiters have actually rejected applicants in light of the private information they found online on social website. Social networking sites provide a platform to organizations where they can encourage applicants to connect as well as enhance the sensibility among people to use these social media platforms wisely. For instance, many of the Step Stone's clients such as Price Water House Coopers, KPMG and the Royal Bank of Scotland are using social networking sites at different points within the recruitment process to assist candidates. Another good example of this has been when organizations found out Facebook groups for university students and interns who have an interest in working with them in near future. Once these groups are created, potential candidates are often made conscious of jobs and opportunities within the organization through specially created blogs and forums. Very often, this is an excellent way of allowing potential candidates to network with existing employees already performing in the company. An alternate approach would involve linking up the talent pools created on the networking sites to the company's e-recruitment portal. This permits organizations to directly outreach to candidates with specific job opportunities, to collect CVs and other personal data with the candidate's permission, and to use them to look for the simplest candidates and link them to the foremost appropriate positions without compromising privacy.

E-recruitment specialists have complete and up-to-date knowledge of knowledge protection laws, to firm sure that they constantly have all of the right policies and procedures in place. This requires a constant contact with all the web candidates, ensuring that they agree with the information that's available about them online and that they are proud of how this data is being judged. There's nothing "unintentional" about the info that's shared between candidate and employer, making this online recruitment process completely legal and above the board. HR staff should view social media sites as channels for building pools of talent and employer brand, instead of free online databases for evaluating candidates. A strategic approach to harnessing social media will help the business attract the candidates it desires and within the limits of prevailing law (Viadeo, 2007).

Applications of AI in Recruitment

Adoption of Artificial Intelligence (AI) has emerged as of the foremost embraced trends among hiring professionals in 2018. Recruiting the suitable talent and managing the range within the workforce has been the formidable challenge. Within the new age and with the facility of AI, the recruitment industry is empowering the employers to satisfy the recruitment challenges. AI is the new building block within the recruitment industry.

Artificial intelligence results in real-time and personalized communication leading to efficiency in candidate acquisition procedure. Promising candidates find it frustrating once they don't hear from recruiters, who sometimes take quite a long time to start the screening of candidates. AI is employed to let the rejected candidates know in time that they can advance with their further job search. AI recruiting software grades and ranks the resumes for skills, experience and other qualifications and responds with an affirmative or rejection message within 24 hours of the receipt of application. With AI rejected candidates are provided feedback concerning the deficiency in qualifications or skills thus aiding candidate to enhance by creating a high-touch positive experience. AI powered systems can nurture candidates by offering positions which may be of interest to candidates for current and future roles. AI assistants engage candidates through the online, mobile, or social platforms. AI assistants intelligently provide next steps and may route suitable candidates on to recruiters.

Powered with AI and NLP (natural language processing), the AI assistant acts as a front face for candidate communication. AI assistants allow recruiters to specialise in what they're best doing at, while AI assistant focuses on candidate capture, screening, scheduling, coordination, communication and engagement (Paradox, 2018). Chat bots are AI-powered assistants that enable real-time and private engagement with the candidates. Candidates interact with these AI-powered assistants through text messages, emails or a dialogue box. For improving the recruitment process, the AI assistant provides feedback on the candidate experience to the recruiter. AI is intelligently programmed to avoid unconscious bias. AI-powered systems can ignore primary sources of bias like names, schools attended, gender, age and race (Bullhorn, 2018).

With AI, recruiters can process volumes of data to seek out suitable candidates. AI can scan social media data to assess the candidate's values, beliefs and attitudes. With AI, recruiters are getting clues on personality traits and suitability beyond the normal resumes. AI is unbiased and screens resume fairly by giving equal weightage to all candidates. AI is disrupting the recruitment industry and is replacing repetitive tasks that were traditionally performed by human recruiters. AI is making some processes obsolete, as recruiters are happy to delegate the mundane, repetitive tasks to AI-powered systems. AI is changing the roles that the recruiters play and is resulting in more thoughtful hiring. With AI taking care of boring and repetitive tasks, recruiters can now focus on becoming creative and may specialise in strategic issues. Recruiters now have time for long-term planning and building personal relationships with existing employees and new hires.

AI is swiftt at identifying talent but still the activities like rapport building, assessment of cultural fit and negotiation must be done by humans who now act as talent advisors. AI is speeding up the hiring process. AI is making the connection between man and machine more productive. Human as recruiters are still needed to know, interpret and quiz the candidates for the proper mixture of empathy and emotions. AI tools got to be programmed with the proper parameters to place a barricade to the unconscious bias. Recruiters got to understand and master the software program and therefore the results. Modern recruiters should communicate a positive image of the organization as creative, innovative and therefore the best place to work with. AI helps in speeding up the recruitment process. AI-powered Bots connect to candidates post submission of application, solve queries and have interaction with screened candidates during the hiring process.

Applicant Tracking System (ATS)

Many businesses use applicant tracking systems (ATS), also referred as talent management systems, to process job applications and manage the hiring process. It supplies an automatic way to companies for managing the whole recruiting process, from receiving applications to hiring employees. The knowledge within the database is employed to screen candidates, test applicants, schedule interviews, manage the hiring process, check references, and complete new-hire paperwork. When applicants apply for employment online, their contact information, experience, educational background, resume, and canopy letter are uploaded into the database. The knowledge is then transferred from one component of the system to another as candidates move through the hiring process. The system framework allows company recruiters to review the applications, send applicants automated messages letting them know their applications are received, and provides online tests. Hiring managers can schedule interviews and mail rejection letters through the ATS. Finally, human resources personnel can use equivalent information to place individuals on the payroll once they're hired. These integrated systems streamline the whole recruitment and hiring process for employers.

Using an ATS saves both time and money. Information from applicants is uploaded and arranged during a database, making it easily accessible and searchable for human resources professionals. The information is collected and automatically organized digitally, companies need not to pay overtime to sort and file paper applications.

Some systems can also save job applicants' time. Many employers use systems that allow job applicants to upload their vital information, work histories, education, and references directly from their profiles on websites like LinkedIn or Indeed. While job applicants got to customize their application materials for various positions, having the ability to bypass the tedious process of retyping this information for each application may be a valuable time saver (Doyle, 2020). According to Costa (2020), in today's job market, companies that use an Applicant Tracking System (ATS) are most likely to quickly attract, identify and hire the best candidates. On the other hand, those who do not leverage new recruitment strategies and tools are certain to fail to achieve recruitment objectives. Driving an efficient recruitment strategy highly depends on the ability of a company to implement more streamlined and agile processes that enable recruiters to focus and spend more time on what matters the most i.e. engaging with top talent. Digitizing and centralizing all the candidates' information in a single platform and performing an initial screening of the incoming applications to signal the best prospective candidates are just some of the things an ATS can do for businesses. Applicant Tracking Systems can improve the entire recruitment process and ease up some of the most important pain points regarding hiring and talent acquisition activities. some of the advantages are highlighted below:

- **Reduces Time Spent with Administrative Tasks:** ATS software allows organizations to accelerate their recruitment process and spare a considerable amount of professional time for more important tasks. It helps in eliminating the manual publication or listing of jobs on job boards, one at a time, or sending follow up emails individually. An ATS allows recruiters to do all that with just a few clicks.

- **Provides Better and Faster CV Screening**: Everyone agrees that it is tedious for recruiters to manually screen every incoming application for a new opening, given that each job offer receives on average 250 applications. It is a heavy work load for a handful of recruiting executives to screen and arrange face-to-face interviews. An ATS automates the initial screening of all applications, using algorithms to match candidates against job requirements. People who don't meet minimum requirements are rejected and therefore the rest is ranked consistent with qualification indicators. This allows recruiters to stop wasting time with unsuitable applicants and getting tired to correctly identify promising profiles.

- **Facilitates Collaborative Hiring:** Applicant Tracking Systems offer amazing features that facilitate collaborative hiring among companies. HR teams can easily and quickly share notes and ratings on every candidate and drive a collective decision-making process, in which every recruiter has a say.

93

Collaborative hiring has major benefits for companies, such as assembling more diverse and successful teams.

- **Improves the Quality of Hire:** The quality of hire is also improved using Applicant Tracking Systems. Moreover, there are many reasons for that to happen. For instance, thanks to the initial automated matching of candidates, recruiters only engage with the best candidates within the applicants' pool. Also, with the automation of administrative tasks, recruiters are longer available to interact with candidates within the selection process. This helps in gathering more information and make detailed assessments of the candidates before making a hiring decision.

- **Speeds up the Recruitment Cycle:** Having all candidates' information centralized and organized in one single platform, gives recruiters faster access to the hiring process overview. Automated communication features also increase the recruitment cycle, making it quicker to provide general follow up and to fast move candidates for the next steps. Overall, an Applicant Tracking System allows recruiters to develop and execute a more efficient workflow.

- **Boosts Employer Brand:** A consistent and interesting recruitment process strengthens and boosts the employer branding. With the assistance of an Applicant Tracking System, companies can attract and have interaction with more candidates. Features like a complete branded Career Pages, for instance, allow impressing candidates and improving organization's talent pipeline.

- **Enhances Candidate Experience:** Besides building a more efficient hiring process, Applicant Tracking Systems enhance the candidate experience, which is important for the success of hiring process. After all, satisfied candidates become the future brand ambassadors. There is no doubt that companies need to adapt their strategy and implement tech solutions that enable better and faster hiring.

- **Privacy Concerns:** The rapid and partly unlimited communication over social media could cause a breach of confidentiality by making participants' data public. Social media platforms like Facebook and Instagram allow users to comment directly on advertisements. Comments could reduce privacy as participants may be discussing, inquiring about specific job or recruiter's information by tagging friends that they may know personally. Additionally, comments on advertisements could be negative and deter people from joining the firm.

CHALLENGES OF SOCIAL MEDIA RECRUITMENT

This chapter has covered a number of opportunities and challenges human resource management could face when use social media at workplace. The opportunities that HRM can achieve when using social media might enhance recruitment techniques, communication and productivity of the organization. Additionally, using social media can reduce training cost since everything could be implemented online such as training, classes and discussions. Furthermore, using social media might prevent HR recruiters to get into any lawful dispute, due to discrimination and/or negligent when they search for information about candidate before interview. Furthermore, social media is a good source where management may perhaps obtain feedback about people who are interested in the company and getting information about the performance of specific tasks in the organization. Social media carries multiple opportunities as well suffers with many challenges such as candidates waste their valuable time surfing in the social surfing instead of working with proper attention in the tasks and responsibilities. Despite sometime social media can prevent recruiter from legal dispute; however wrong recruits through social media could lead to serious issue in the organization. Furthermore, unsatisfied employee could use social media platforms such as company blogs as a tool for taking revenge from their seniors or employer which will not only destroy employer's brand but the whole organizational reputation too.

In addition, growth of social media technologies putting a new turn to employment problems, since employees gradually access more to public media, the recruiter have been showing a lot of interest to regulate and to monitor what the candidates convey online. Besides, using social media could cause higher security risk for organization information and data because of malware or spyware. Sometime it could cost company's money and energy in preparing older generations to use social media platforms for communication. Additionally, HR face challenge to balance between using social media and traditional method of recruitment, because, only focusing on social media; they maybe excluding qualified staff who have no access to the social networks or who already choose to desert the use social networking technology for applying for job. Final challenge in HRM using social media is that to keep up with dramatic enhancement in technology and social media.

Presently there are a lot of corporations, agencies and companies who generally tend to apply social media channels and the structures for recruiting candidates. The primary cause the agencies are recruiting via social platform is the cost effectiveness and time saving exceptional of the useful resource. While social media plays multiple roles, recruiters are using it as an economical way to hire personnel. LinkedIn, Facebook and Twitter are within the top lists of the recruiters for posting Job vacancies. Social media for recruiting are often described as an intersection of recruitment and social media. While social recruitment gives an oversized range of advantages, it also has some negative aspects that the consumer have to recollect of, just so he may also be prepared to encounter any negative brand picture (Baker, 2018).

Automation is the foremost concern for the industry and is ranked among the highest three challenges for the recruitment agencies within the UK. Interestingly despite all the mention automation in recruitment from semantic search and chat bots to online on-boarding, adoption is that the biggest challenge and is a smaller amount than or adequate to 11 per cent for firms which are using it to an excellent extent. Within the UK recruitment industry, for functions like screening or credentialing, many organizations are not using any automation and one-third of the companies are not using any technology for selection, on-boarding or candidate nurturing (Bullhorn, 2018).

Due to the rise in standards round the worldwide media also provides hard and difficult demanding situations even as using the social media within the system of recruitment. Examples encompass laws concerning personal area, just like the European Union's General Data Protection Regulation (GDPR). "With GDPR, there is much less bang in your dollar in online social media marketing, as an advertiser, than there was a few months ago. This will stage out as companies figure out how to work with the GDPR policies, but for now, it's far a hurdle to look out for," Musto warned. Others are legal guidelines governing honest hiring practices, like folks that forbid discrimination supported age, race and different elements which can be generally located on social media profiles. "Recruiters ought to remember of federal and state laws which have positioned restriction on permission of expertise for employers. The firms or the business aren't at all equipped to request enrollments for his or her profiles which they use on the social media," stated Marco Piovesan, CEO of InfoMart, a worldwide heritage screening organisation.

FUTURE OF SOCIAL MEDIA RECRUITMENT

There is no doubt the fact that there is advancement in the technology which also has changed the entire outlook of how we see the world. It created an entire global platform and is now entering new markets on daily basis. The use of technology has increased day by day and professionals of different sectors are utilizing various techniques to advance and progress in their ways. The advent of technology is also helped recruiters and HR agencies. Use of the Technological advancements has revolutionized the way HR managers and recruiting directors perform their tasks and jobs. The whole recruitment industry, the old-school techniques that were used by the recruiters have changed with the coming of the Internet and the technologies. Finding and hiring ideal candidates is a tough as well as expensive job. The recruiting landscape is changing every single day, and to look for the best employees, it is crucial to have an appropriate recruiting approach. The recruiting efficiency, revising and making changes in existing plan have increased. Recruitment marketing has emerged as one of the top trends of 2019. It has now turned into the combination of strategies and tactics used by an organization to attract and engage the job seekers. The main aim of having a recruitment strategy is to attract the top candidates to fill the open position by making it more appealing for the candidates. There are many candidates and job seekers who are using mobile technologies to apply for various jobs. Companies are using social media to assess the soft skills of the candidates. Apart from paper resume, the companies have started evaluating the digital portfolios of the candidate. Social awareness and visual recording submissions has created a perfect and efficient impression. The recruiters are able to visualize whether the applicants are suitable for job role or not.

The entire social life is looking forward for digital world; the online interactions with the candidates for their interviews and resumes are becoming more structured and robust. Hence, now as a job seeker, one should start building up digital portfolio and create an innovative video resume. Artificial intelligence is replacing many tasks, but it is true that it will not replace all jobs. AI has influenced many industries, and one of them is the job market. AI has become a must-have in the recruiter's toolbox, and the recruiters have started working with the assistance of AI. In 2018, the use of AI in recruitment was in abundance whereas in 2019, the trend has drastically increased and it is time to implement it. HireVue and Knockri are the artificial intelligence software which has helped to detect and assess the stressfulness of the employees with the help of the structural language. These technologies have also helped in determining if the candidates are answering the questions truthfully or not.Big companies such as PepsiCo have also started using AI to conduct their interviews. They use automated video messages and phone calls. The system also provides enough time for the candidates to give answers to each question.

Gamification is one of the most surprising developments in the process of recruiting. It is a viral new trend and is the most suitable ways to decide if the member is right fit or not. It involves using the elements of game and then joining these with other non-gaming structures some of them are business world and the corporate world. Organizations are now using these principles to examine the candidate's reactions in and under specific environment and several pressures. Another exciting development that has been discovered is the use of virtual reality in the onboarding process. Incorporating virtual reality will improve the interview process and will also increase hiring efficiency. The use of virtual reality will give a better picture of how the candidates will respond in the practical settings.

The role of HR and recruiters has changed drastically over the past few years and will continue to improve in the future which is why it is crucial to keep an eye on the latest trends. Applying the trends effectively with the right methods will take recruiting strategy to another level (Aditya.S, 2019).

SKILL-SETS REQUIRED IN RECRUITERS

Recruiting is on the rise, but it is also changing. Companies need Recruiters. They are one of the most in-demand jobs as businesses brace for a future where talent is everything. The team will need to master new skills, metrics, and tools. As automation progresses, jobs are getting less repetitive and more creative. That does not just mean jobs that are more interesting, it also means that talent will make a bigger business impact than before. Companies are hiring more recruiting professionals right now than at any other time in the last five years, according to LinkedIn data. It presents that companies around the world are looking to strongly build up their talent acquisition teams. The future of recruiting will be fiercely competitive—even when it comes to hiring and retaining the organization's recruiting team. The recruiters will need to start capacity planning. With demand for recruiters on the rise, the companies should take stock now and see if there is need to hire more recruiters to meet company's future hiring needs. Learn how LinkedIn's talent acquisition team predicted hiring demand and staffed recruiters accordingly. Majority of the organizations and consultancy firms have agreed that retaining top recruiters would be a major priority over the next 5 years. Given how hot the hiring market for recruiters is, there may be need to rethink and re-incentivize the policies to retain recruiters. Recruiters need to build a strong business case and should have the business's attention and take advantage of that. One should be good to create a plan to support the company's hiring needs over the next 5 years with the given resources.

Recruiting is not just important today—it is much more difficult. That is because companies' talent needs seem to change day by day. The top priority for recruiting organization over the next five years will be keeping pace with the changing technology and company's rapidly changing hiring needs. To keep up, the recruiters will have to stay close to the businesses by aligning with leaders very often. They also have to hire flexible people in recruiting teams who are better able to adapt to new needs quickly. It is one thing to execute a plan; however, most important is to design the most effective plan. More and more companies are looking for recruiters to do both. For a long time, many hiring managers treated recruiters as order takers. In the future, recruiting will surely be recognized in a strategic partner role. Recruiting leaders and recruiters themselves will be expected to bring a perspective, push back, and lead the way forward. That means, not only aligning with business goals, but also advising leaders on the best way to achieve them. As the most administrative and routine parts of work get automated, jobs are getting more creative and complex which applies to recruiters as well. The function will become less about execution and more about talent strategy. Recruiters will have to anticipate needs, solve problems, and spot opportunities for hiring managers and their teams. In the future, recruiters will look more and more like HR business partners as stated by Tristan Klotsch, Vice President of HR at Serrala. Results-based metrics will rule the future. Many teams are not there yet, but could be soon. The two most impactful metrics of the future are both strategic and results-based measures. Quality of hire and sourcing channel effectiveness—a measure of which sources produce quality hires, both account for the business impact of the people the team is bringing in. Most recruiting teams still are not tracking either, but the metric with biggest gap between usefulness and actual use is candidate experience, a measure that companies expect will become more commonplace in the near future.

CONCLUSION

Nowadays, only the technological changes are happening so quickly, developing with society changes and even changes in community values and these trends in many ways have become equally ingrained. Many people have thought that the utilization of the social networks in HR will simply disappear with time like other resources. To present there is no substitute of social media, which provide the best of technology and platforms to keep in touch with employees, but then there is nothing better than being there. In the world of social media, organizations are becoming unidentified, where any recruiter can recruit to any potential candidate. The last decade has provided the HR management with social media, which is providing enormous reach as a deep sea for better dives. However, at the same time, organizations and recruiters should not forget the importance of human touch and relations as the candidates might have fake accounts which can negatively impact companies in long run. HR faces significant challenges as they form strategies that will enable them to take advantage of the opportunities inherent in this dynamic and emerging informative century. Therefore, established organizations that follow quickness and innovation will do well to employ this new generation through social media, however, for traditional companies, such appointments are both a serious challenge and the important opportunity.

REFERENCES

Aditya, S. (2019, June). *6 latest recruitment trends to implement in 2019*. Retrieved from https://www.zippia.com/employer/6-latest-recruitment-trends-to-implement-in-2019/

Allden, N., & Harris, L. (2013). Building a positive candidate experience: Towards a networked model of e-recruitment. *The Journal of Business Strategy, 34*(5), 36–47. doi:10.1108/JBS-11-2012-0072

Armstrong, M. (2006). *A handbook of human resource management practice* (10th ed.). Kogan Page.

Arpit, M. (2018, January 2). *A brief history of recruitment and hiring*. Retrieved January 2, 2020, from https://www.hackerearth.com/blog/talent-assessment/brief-history-recruitment/

Baker, P. (2018). *Recruiting with social media: Tips, tools and challenges*. Academic Press.

Broughton, A., Foley, B., Ledermaier, S., & Cox, A. (2013). *The use of social media in the recruitment process*. Retrieved December 20, 2019, from http://www.acas.org.uk/ media/pdf/0/b/The-use-of-social-media-in-the-recruitment-process.pdf

Brown, V. R., & Vaughn, E. D. (2011). The writing on the (Facebook) wall: The use of social networking sites in hiring decisions. *Journal of Business and Psychology, 26*(2), 219–225. doi:10.100710869-011-9221-x

Bullhorn. (2014). *Global social recruiting activity report*. Retrieved January 2, 2020, from https://www.bullhorn.com/resources/2014-social-recruiting-activity-report/

Bullhorn. (2018). *2018 UK recruitment trends report: the industry's outlook for 2018*. Retrieved January 2, 2020, from http:// pages.bullhorn.com/rs/131-YQK-568/images/2018%20Trends%20Report_UK.pdf

Business Tech. (2016). *Massive rise in Facebook users in South Africa*. Retrieved January 3, 2020, from https://businesstech.co.za/news/general/128584/massive-rise-infacebook-users-in-south-africa/

Caers, R., & Castelyns, V. (2011). LinkedIn and Facebook in Belgium: The influences and biases of social network sites in recruitment and selection procedures. *Social Science Computer Review, 29*(4), 437–448. doi:10.1177/0894439310386567

Chaudhary, M. (2017). *LinkedIn by the numbers: 2017 Statistics*. Retrieved January 2, 2020, from https://www.linkedin.com/pulse/linkedin-numbers-2017-statisticsmeenakshi-chaudhary

Costa, D. (2018). *Applicant tracking systems: main advantages of using an ATS*. Retrieved January 2, 2020, from https://www.skeeled.com/blog/applicant-tracking-system-main-advantages-of-using-an-ats/

Darrow, B. (2017). *LinkedIn claims half a billion users*. Retrieved January 2, 2020, from https://fortune.com/2017/04/24/linkedin-users/

Doherty, R. (2010). Getting social with recruitment. *Strategic HR Review*, *9*(6), 11–15. doi:10.1108/14754391011078063

Doyle, A. (2020). *How employers use applicant tracking systems (ATS)*. Retrieved January 5, 2020, from https://www.thebalancecareers.com/what-is-an-applicant-tracking-systems-ats-2061926

Dutta, D. (2014). Tweet your tune – Social media, the new pied piper in talent acquisition. *The Journal of Decision Makers*, *39*(3), 93–104. doi:10.1177/0256090920140307

Engstrom, E. (2015). *Evolution of recruiting*. Retrieved from https://recruiterbox.com/blog/evolution-of-recruiting-infographic

Galanaki, E. (2002). The decision to recruit online: A descriptive study. *Career Development International*, *7*(4), 243–251. doi:10.1108/13620430210431325

Hoffman, D. L., & Fodor, M. (2010). Can you measure the ROI of your social media marketing? *MIT Sloan Management Review*, *52*(1), 41–49. Retrieved January 2, 2020, from http://www.mitsmr-ezine.com/mitsmriphone11/fall2010/m2/Mobile Article.action?articleId=23732&mobileWeb=true&lm=1285614348000%5Cnh ttp://www.emarketingtravel.net/resources/can you mesur the ROI of your Social media marketing.pdf%5Cnhttp://sloanreview

Houran, J. (2017). *New HR study: Candid recruitment experiences with LinkedIn*. Retrieved January 2, 2020, from https://www.aethoscg.com/aethos_insights/newhr-study-candid-recruitment-experiences-with-linkedin/

Jindal, P., & Shaikh, M. (2014). Social networking sites – Emerging as effective tools for attracting talent. *Gavesana Journal of Management*, *6*(2), 48–55.

Jobvite. (2014). Social recruiting survey results. *Jobvite*, 1–17. Retrieved January 2, 2020, from https://www.jobvite.com/wp-content/uploads/2014/10/Jobvite_ SocialRecruiting_ Survey2014.pdf

Joos, J. G. (2008). Social media: New frontiers in hiring and recruiting. *Employment Relations Today*, *35*(1), 51–59. doi:10.1002/ert.20188

Khullar, A., Pandey, P., & Read, M. (2014). Effective use of social media recruiting. *International Journal of Management*, *4*(4), 216–227. Retrieved January 2, 2020, from https://www.inderscienceonline.com/

Lennox. E. (2017). *The history of recruitment.* Retrieved January 2, 2020, from https://recruitingblogs.com/profiles/blogs/the-history of recruitment

Melanthiou, Y., Pavlou, F., & Constantinou, E. (2015). The use of social network sites as an e-recruitment tool. *Journal of Transnational Management*, *20*(1), 31–49. doi:10.1080/15475778.2015.998141

Nikolaou, I. (2014). Social networking web sites in job search and employee recruitment. *International Journal of Selection and Assessment*, *22*(2), 179–189. doi:10.1111/ijsa.12067

Ollington, N., Gibb, J., Harcourt, M., & Doherty, R. (2013). Online social networks: An emergent recruiter tool for attracting and screening. *Personnel Review*, *42*(3), 248–265. doi:10.1108/00483481311320390

Paradox. (2018). *AI reinvents the candidate experience*. Available at: https://paradox. ai/

Parez, M., Silva, K., Harvey, D., & Bosco, S. (2013). Linked into a job? The ethical considerations of recruiting through LinkedIn. *Proceedings for the Northeast Region Decision Sciences Institute (NEDSI).* Retrieved April 20, 2017, from http:// docs. rwu.edu/cgi/viewcontent.cgi?article=1000&context=management_theses

Parker, M. (2008). Can social networking sites be recruitment tools? *Strategic HR Review*, *7*(3), shr.2008.37207caf.003. Advance online publication. doi:10.1108hr. 2008.37207caf.003

Phillips, J., & Gully, S. M. (2012). *Strategic staffing* (3rd ed.). Pearson Education.

Safko, L., & Brake, D. K. (2009). *The social media bible: Tactics, tools, and strategies for business success. In Skin & Allergy News* (Vol. 43). John Wiley & Sons.

Singh, K., & Sharma, S. (2014). Effective use of social media for talent acquisition and recruitment. *International Journal of Intercultural Information Management*, *4*(4), 228–237. doi:10.1504/IJIIM.2014.067932

Sinha, V., & Thaly, P. (2013). A review on changing trend of recruitment practice to enhance the quality of hiring in global organizations. *Management, 18*(2), 141–156. Retrieved April 20, 2017, from https://hrcak.srce.hr/index.php?show=clanak&id_clanak_jezik=166315

The future of recruiting report. (n.d.). Retrieved from https://business.linkedin.com/content/dam/me/business/en-us/talent-solutions/resources/pdfs/future-of-recruiting-report.pdf

Zephoria Digital Marketing. (2017). *The top 20 valuable Facebook statistics – Updated July 2017*. Retrieved April 20, 2017, from https://zephoria.com/top-15-valuablefacebook-statistics/

Zide, J., Elman, B., & Shahani-Dennig, C. (2014). LinkedIn and recruitment: How profiles differ across occupations. *Employee Relations, 36*(5), 583–604. doi:10.1108/ER-07-2013-0086

Chapter 6
Identifying Innovations in Human Resources:
Academia and Industry Perspectives

Amrik Singh
iD https://orcid.org/0000-0003-3598-8787
Lovely Professional University, Punjab, India

Sanjeev Kumar
Lovely Professional University, Punjab, India

ABSTRACT

HR is evolving into a more technology-based profession because organizations needs to streamline HR processes and reduce administrative burden, reducing administrative cost; compete more effectively with global talent; improve services and access data to the employees and managers; provide real-time metrics in order tom on spot decisions for the decision makers; and manage the workforce more effectively and enable the HR to transform so it can play more strategic role in the business and operations. The purpose of this chapter is to develop a meaningful debate on the innovations in human resource in terms of new ideas, methods, and technology to better meet the evolving requirement of the organization and workforce. Anticipating and exploring the future needs and circumstances rather than simply finding some responses to the situation, this chapter highlights challenges and prospects related to innovations in HR.

DOI: 10.4018/978-1-7998-4180-7.ch006

INTRODUCTION

To maintained the long relationship and survive in the globe new innovations should bring in the organization for the development of new culture and traditions. Creativity plays an important role of competitive advantage of business organization. Researchers suggested, for long term survival organization creativity makes an important role to organizational effectiveness, because it prepare to organization achieve a competitive advantage in a rapidly changing environment. It is the prime duty of HR department to build up healthy relationship with employee in term of creativity and achieve the organizational goal. Hr department using various new tool and software where they can store all the details of employees and give excess to their employee where employee get the information without wasting the time and that time can be utilized in the operations to increase the output of the organization. The local government fulfills the requirements of the society and increasing the value of life, Organization play an important role of creativity bringing life through inventive, customer desires creating job for contributing to the economy (Ambardar & Singh, 2017). HR applies more effective practices and cut down the problems. These are the three dimension of innovative Human Resource Practice (IHRPs). (Ambardar & Singh, 2017; Ruël, Bondarouk, & Looise, 2004). The term Human resource management (HRM) is used widely but defined very slackly. HRM have different policies related to employee benefits and increase the employee commitment, Quality of work, flexibility and organizational integration. Furthermore some of the UK organization applies to this model and many are moving slowly to apply this model for example policies of employee involvement. Robotic Process Automation (RPA) grows up to 200 billion US Dollar in next four year, because more than fortune 1000 companies adopting it (Papageorgiou, 2018) . Human opted for innovation and improvement since the beginning time and progress was slow until the Industrial revolution. When each industry breach was adopted by others, the first attempt at automation was started.

Innovation Tool Used by HRM

According to Homer, "Inspiration is a muse, and innovation seeds can often be found in the most unexpected places." forward to 2018.Today many innovations like information technology (IT), artificial intelligence (AI), machine learning, chatbots and RPA. Innovation speed, variety and volume continuous changing across the world and interact each other professionally. But these changes have not supported to human resources benefits function. In business digitalization process HR is once in lifetime opportunity to be strategic and integrating the business. Changing the workforce to purpose based employment, which is task and skill based employment in past (Geer Jr, Tumblin, & Solomon, 2001; Papageorgiou, 2018; Schraft & Schlaich, 1988). Automation technology depends on use, misuse, disuse and abuse.

Advance Bio-Metric machines use for the marking of attendance which will help to cut the time of HR department which provide accurate data and up to date information. The factor of human use of automation can improve effective training method, system design and judicious policy linking automation use (Parasuraman & Riley, 1997; Sheridan, 1992) . Customers support centers reliance on our agents who acquire skills for that is being delivered. There are many skill requirements in call centre like good speech of that language which is speaking on telephone. Other skill is knowledge of computer i.e. knowledge of keyboard, monitor and internet and other electronic devices. When receiving the call customer support agent speak language well, perform the assign task, giving the right information and speaking sound delightful when speaking customer. A method for screening applicants, the customer is asked, via the company web page, to giving responses to a set of question customized to the screening body and responses to the queries are stored and the process of meeting with customer and support is started on telephone and computer (Schalk, Stovall, & Brooks, 2011).

Challenge Faced by HRD

These are the challenge of twenty-first century management. The first challenge to structure companies that can "change as fast as the world around us" says Hamel and Denial, the lack of capacity to innovation of new options which is planned and shares strongly. The second challenge were hardly distributed to ability of innovation instead of old mental method and inadequate respect for testing, are seen as major barrier. The third challenge are "too much management, too little discretion", "too much hierarchy, too little community" and "too much exhortation, too little purpose." These three challenges are mutually dependent. The third challenge is solving to key the other challenge. A latest survey of 8600 employee in different organization found less than 15 percent of employee engaged in their job and quarter was totally disengaged, prompting the researcher to offer "hierarchy of human capabilities". Value creation in terms of contribution is ascending order these are intellect, creativity, diligence, obedience and passion, these are obedience, diligence, intellect, initiative, creativity and passion. Later it is find out intellect, diligence and obedience, is fast becoming global commodities, while company differentiate capabilities with the greatest potential like initiative, creativity and passion, benefit give or withhold to employee day to day and moment by moment. These process are creating a kind of organizational model that will deserve gifts is critical management challenge. For better results organizational apply new management principal in any organizational model. It is not complete example of future best practices. But this is a leading effort at Whole Food Markets, Google and WL Gore. Food retailing sector is one of the fast growing market in USA. USA company give proposal to customer that people give high price for organic and locally food. It is difficult to find 3 companies to offer these products, In this situation company its own way is a modern management pioneer. Taken together they help to underline how radical deviations from management orthodoxy can have a significant commercial impact.

There are 3 key principal to improve the chance of achieving radical management system by Hamel

- For finding the new long standing management approach that restrict creative thinking to adopting a disciplined approach
- Management give power to apply new idea, and
- Capitalizing on insights drawn from the practices of "positive deviants".

Management is a science; it is defining that "positions us toward new truths." Hamal says "Modern management is based on hierarchy, specialization, planning standardization, control, goal and alignment". These are maximizing operational competence in huge scale organization and this is "modern management is fully competent to address. "When it comes to the challenge of how to create organizations that are highly adaptable and appealing, these philosophies are "insufficient and often toxic." So management look new option? According to Hamel there are five subject and literature i.e life, market, faith, democracy, and cities. Life refers to principal of diversity, from market approach refers to flexibility, democracy built up activism, faith gives meaning and cities approaches furnish providence. New Urbanists thinker like Richard Florida and Jane Jacobs, how great cities contribute to generation of "new pools of economic use" through enabling different man to meet and find out opportunities to trade information, goods and ideas, and how, at least in part, cities are "able to reinvent themselves because they make it easy for individuals to reinvent themselves." Great cities also explain us how to arrange for serendipity. Few have dedicated to reinventing their management process while businesses have worked vast to reinventing their HR process. The exception is General Electric, Procter & Gamble and Whirlpool, but it is true internet, the "most flexible, innovative and engaging thing that human beings have created" and in several ways "the new technology of management" because it distribute the path of creativity widely, ideal allow to race on equal track, decentralize everything, freedom of voice to everyone and encourage resources to follow opportunities (Andresen et al., 2002; Hamel, 2008; Noe, Hollenbeck, Gerhart, & Wright, 2017; Panigrahy & Pradhan, 2015; Papageorgiou, 2018; Schraft & Schlaich, 1988; Walke, 2013). HRM practices power to do financial performance positively; there is a certain reason for expectant them to positive innovation performance. We examine the empirical model through its all hypothesis using 1900 business firms which survey conducted by Danish using principal component analysis. We find out two HRM systems which are conductive to innovative. Researcher examine 9 variables, first variable is equal for the ability to innovative. Second variable is firm internal and firm external training. We have correlate with first system to observing four manufacturing sectors and seconds are five wholesale trade and to the ICT intensive service sectors correlate with the second systems (Laursen & Foss, 2003).

REVIEW OF LITERATURE

Strategic human resource management (SHRM) focuses on to build up strong relationship among individual motivation and their performance.(Schuler, 1992) Various study were conducted on innovation in performance where academia and industry perspective point of view HR expand too much of efforts, time, resources and money. There are number of business theory has been adopting (Jussani, Krakauer, & Polo, 2010). As mentioned in the study that different types of innovation are applicable for different environments (Damanpour & Gopalakrishnan, 1998). In industry point of view HR department create different opportunities and competitive (Chan Kim & Mauborgne, 2005) as per 'Blue Ocean Strategy' it creates different innovation in the field of Academia. In the "Blue Ocean Strategy" which provides new market place, the chance of competition should be less which HY may be explores their idea in term of new innovations which create positive environment in the work place New strategic followed in innovation by the HR give positive results and make organization profitable (Jussani et al., 2010) Invention itself create a positive environment in the organization which have high impact on economic (Serafim, 2011) Today's HR focus on corporate culture in the university and create positive environment among the staff and stakeholders (Araújo & Garcia, 2009) Integration and implementation of Corporate Education e.g promotional activities in the university and their work culture. HR brings various new innovations by giving training and development in the respected areas to meet the objective of the organization to provide positive direction towards the skills, abilities and attitude. (EBOLI, 2006) There are number of new applications of innovation Technologies introduced by the human resources management which can improve the quality and efficiency of the HR department. By introducing the new innovation practices in the academia and industry must get some advance internal and external information with their unique features in term of technologies (García-Carbonell, Martín-Alcázar, & Sánchez-Gardey, 2015) HRD focus on education sector by implementation of new trends followed by the Human Resource in term of recruitment of new staff and their orientation programme. Human Resources Management (HRM) performed various function in term of recruitment, training, remuneration and compensation, and legal issues (Joseph & Ezzedeen, 2009) Human resource management focus on modern techniques and latest trends in HR management practices for better results and performance system (Thom, Ritz, & Masiulis, 2004) A tremendous change has seen in the technology development and their smooth operations in globally. e-HRM's concepts introduced in the academia and industries which will help to reducing the number of task, solving the problems, cut down the administrative cost and increase the efficiency of the work and bring the quality work where manager and employees become fully satisfied.(Analoui, 2007) "Improving organizational performance, managers and decision makers should make their HRM systems more visible, understandable, legitimate and relevant"(Katou, 2015) HR experts believes that it fulfill the needs of employees and these new innovation can also helps in administrative efficiency work, knowledge improving and better results which can build up the confidents (García-Carbonell et al., 2015) Researcher focus on the benefits of e-HRM (Ruël et al., 2004) improving the orientation of HRM, cut down the daily cost, increase the effectiveness, improvement in services and reduce the pressure of managers and employers.(Kavanagh & Johnson, 2017) The key criteria that must be taken in to account for successful innovation through business environment and capacity building analysis (Navin, Navimipour, Rahmani, & Hosseinzadeh, 2014)

In the globalization era, HR must be artistic and promote the culture of creativity by tapping the competency. Different studies proved that creativity makes an important contribution to increase the effectiveness of the organization. Human Resource act as a middleman to facilitate and communicate and achieve the objectives of organization. In a dynamic environment with rapid globalization and advances in science & technology, "creativity" & "innovation" play an important role for long-term development and sustainability (Vveinhardt & Andriukaitiene, 2016) Innovation requires managing flexibility-control tensions. Today innovation found one of the important significant tools which increase the efficiency of the work (Panigrahy & Pradhan, 2015)

Employee behavior leads the organization in the positive direction and to show the involvement in day to day activities (Hartmann, 2006) Organizational culture consider one of the major component which can stimulate the behavior of employee in the organization (Nacinovic, Galetic, & Cavlek, 2009) Creative Innovation help the supportive cultures which will fulfill the expectation of employee and their moral (Khazanchi, Lewis, & Boyer, 2007) "The intentional introduction and application of ideas, processes, products or procedures which are new and benefit to the job, the work team or the organization" (Ying, 2006) innovation have different steps which includes different idea in term of evaluation, development, tasks orientation and their growth (Panigrahy & Pradhan, 2015) Creativity in the innovation provide new direction to organization which fulfill demand of the employer (Woodman, Sawyer, &

Griffin, 1993). HR automation always facilitates the processes and procedure in any organization. Every organization use variety of software such as ECM (Enterprise Content Management) which reduces the time taken for completion of tasks. An organization uses this software to capture, store, retrieve and secure information. Followings are the some HR tasks which are rapidly moving toward automated:

1. **Employes Record Management:** Retaining employees records as per government regulations in the depository.
2. **Employee Recruitment:** Automatically store application submitted through online forms into ECM repository and assigning them to a recruiter for review.
3. **Employee Onboarding:** Send some confidentiality agreements, waivers and other forms to new hires and once completed send them to corresponding folder.
4. **Benefits:** Track when employees become eligible for benefits.
5. **Health and Safety:** Sending emails to floor monitors when employees leave the organization or move to the other floor.
6. **Tax Forms:** Facilitate the distribution of W2s employees and other tax documents with employees email reminders.

Increasingly technology has a profound impact on HRM. As technology evolves it will also reforms to take new contours in both its processes and practices. HRIS also emerged as in response to the need of change to be carried out in most fruitful way considering the improved accuracy the quick access to information, the increased competitiveness and efficiency and re-engineer of the HR functions. In this fast changing competitive globalized market place innovation has become the essential factor for any organization to achieve success.

THE CHALLENGES AHEAD

Changing Role of HR

To meet the requirements of the organization the HR professionals need to play variety of roles. Traditionally the HR department had limited involvement in the organizational affairs and goals. The functioning is limited with making staffing plans, providing job training programs, running appraisal programs and payrolls. They only focus on the short term and day to day needs of the human resources. But this situation changed and the growing importance of the HR function to the success of the business made them to be more involved in the business process. And the roles of HR department increased the involvement to the longer term and strategic directions of the organization. The changing roles of HRM are like, strategic role, monitoring role, innovator role, facilitator role, enabler role etc (Richard Wolfe, Patrick M . Wright, And Dennis L . Smart (2006); Rosalie L. Tung, Yongsun Paik, Johngseok Bae (2011); Rothwell, R. (1992); Rousseau, D.M. and Wade-Benzoni, K.A. (1994)

Strategic Role

The HR department and their professionals will get involved in the broader decision making process which will provide overall direction about the organization. And they need to understand the business direction in clear and act accordingly in order to achieve the strategic HRM. SHRM is proactive management technique for the people. This requires thinking ahead, and planning the ways for the organization to meet the needs of the employees, and for the employees to meet the needs of the organization. This can affect the way of HR department functioning by improving everything from hiring practices and employee training programs to assessment techniques and discipline (Richard et. al, 2006; Rosalie, 2011)

Monitoring Role

Reviewing and evaluating the strategic plan implemented by the organization to enhance the organizational capabilities will be the major role to the HR department in future. Were there will be more responsibilities to them regarding strategic monitoring and evaluate the process and diagnose the problems in it and determine the reasons for deficiencies. Then revised action plans with all changes will be implemented.

Innovator Role

The organizations are asking their HR department for innovative approaches and solutions to improve productivity and the quality of work life of the employees in order to overcome all uncertainties in the work place. In this changing world innovation become a necessity for all the organization to achieve their competitive advantage. HR departments face demands same as their organizations. In order to achieve their success they must continually update their operations and redesign the work environment. The HR department review & evaluate the expenses then implements incremental changes to become efficient and stay lean. Flexible HR departments forcefully seek to be liberated and setting an example for other departments and line organizations.

Facilitator Role

It is necessary for the organizations to adopt new technologies, change in organizational structures, business processes, work cultures, and procedures to meet the demands of the customers. HR department has the responsibility to provide skilled labors to facilitate organizational change, and maintain organizational flexibility and adaptability. The HR department plays a significant role in organizational change. And they should guide the discussion, flow of knowledge, information and learning throughout the organization in order to achieve success.

Enabler Role

HR policies and procedures are fully realized by the professional and they will act as the enabler to the employees and helps them to acquire knowledge and skills about the new technologies and processes, so that the employees can easily adapt to the change and make themselves more capable towards the organizational capabilities.

- **The War for Talent:** The talent will be corporate resource in future. Smart, sophisticate and technologically sharp employees will be top priority of any organisation. The traditional work force planning will be replaced by the talent strategies and the skill gap analysis. When the gap analysis is made then the HR department will have knowledge about the need of training and accordingly training can be given to the employees in order to enhance them technically efficient. And the HR department will follow the recruitment strategies like employment branding, nurturing relationships, referrals, competency fit etc. to acquire talented employees to the organization.

Future HR Innovations

In the globalisation era, to compete and sustain in a long run, a business establishment must be creative and foster the culture of creativity by tapping the competency of human resources. And in this business of uncertainty, risk and volatility, creativity plays an important function towards creating a competitive advantage for organizations (Panigrahy, Nrusingh & Pradhan, Rabindra. (2015).

- **Outsourcing of HR Functions:** The HR professionals are expected to deliver values in areas like strategic compensation activities, succession planning for employees, talent acquisition, risk mitigation, employee effectiveness and these are the key expectations by the top management. The reasons behind HR outsourcing is to reduce cost, focus more on the organizational functioning, regulatory compliance (legal risks are transferred to outsourcer and helps in obtaining expertise in specialized regulatory), access to best technologies and scarce of internal resources.

- **Healthy Workplace:** There is a link between work environment, employee's health and well being. When the employees are sick and stressed out the organization cannot achieve its competitive advantage. The goal of healthy workplace development is not only for employees but also for the organizational health and success. To achieve healthy workplace development the essentials drivers are, healthy leadership, planning the actions, employee focus and healthy outcomes. And the organization should focus on the employee's health in order to improve their performance the main factors to be considered are, physical, emotional, spiritual, mental and social feelings of the employees.

Diverse Workforce

Diversity in workplace has a reputation for acceptances of employees were they are different in religions, culture, beliefs, languages, customs and traditions. Diversity in workplace is a business strategy. Were the futures of workplace will be complex collection of employees and all with different needs and wants. Diversity is good because an organization with a broad variety of people with a diverse range of perspectives is better able to do business with a variety of people, to solve a variety of problems and to make a variety of decisions.

Technology Driven

Technology provides a great impact on the personal and professional lives. The technology is necessary for all the organization which travels towards success and those measures should reduce employee resistance to new technology and processes and ensure that steps are taken to provide support and education of the staff to cope with the underlying changes. The future of HRM will have drastic change in the process and approach to it. The concept of HRIS (Human resource information system) will be vanished and the organizations will develop their information system to improved employee relationship management.

Leadership Development

Leadership styles are not built through courses. It is an in born quality of the people which qualifies their character. But it needs some process to be followed for further development. Through "action learning" we can develop the leadership among the employees. Action leadership involves group of executives from various background who has strategic interests to solve the issues in the organization. Through enhancing the leadership among employees it helps in encouraging them and rewarding in risk taking. The vertical development (earned through individuals) should be focused well in order to increase their leadership efficiency. The challenges for organizations that wish to increase the vertical development of their leaders and cultures. And it helps in implementing the developmental concepts in the workplace.

Succession Planning

Succession planning will be focused more in future to identify and develop the internal employees to the top leadership positions of the organizations. It helps in increase the availability of the experienced and capable employees to the organization to achieve success. In simple terms succession planning will be focused on talent management concept which helps in identifying talented people for the job. The HR department will change to talent department in future which focus on getting young and new talents to the organization.

HRM Innovation in Context to Academia

The success of any educational institution is believed to rely mainly on the quality of its human resources and its consideration of human resource management as the heart of the educational administration (Jones & Walters, 1994). Higher education is an instrument for improving the social life of a nation. The quality of a civilization depends basically on the character of human being not on the physical equipment or the political machinery. The main task of education, especially higher education, is the improvement of this character. Higher education institutions have become more interested in implementing human resource management as a full strategic partner in their operations. Work life Report (1994) listed some factors that make human resource management a successful strategic partner. Some of the innovative trends identified in educational or higher institutions are:

1. **Employee Benefits:** Takes into account the benefits associated with health, dental, prescription medications, workers compensation, and other benefits related to the wellbeing of employees.
2. **Diversity and Respectful Workplace:** Includes policies, programs, and activities that promote a harmonious environment in the workplace, and show respect for individuals and their roles at the institution regardless of their distinguishing characteristics.

3. **Global Human Resources:** Complying with the rules and laws of the U.S. Immigration and Customs Enforcement Agency, as well as those in any country where an ex-patriot may reside.
4. **Human Resource Management:** Includes practical policies and processes on applicant selection and recruitment, development and training, employee relations, general management and records retention, and legal factors.
5. **Performance Metrics:** Includes comprehensive range of metrics in main human resource programme areas where data should be collected and analysed to explore trends and performance measures.
6. **Recruitment/Selection/Termination:** Includes procedures for acquiring, interviewing, and recruiting of quality employees, in addition to assuring minority recruiting. Moreover, procedures and policies for terminating the dismissal of employees (Jones & Walters, 1994).
7. **Risk Management, Safety, and Health:** Includes providing advices in occupational health, environmental protection, the areas of safety and risk management.
8. **Wage and Salary Administration:** Includes developing and adopting criteria for regulating compensation in a reasonable equitable manner (Jones & Walters, 1994).
9. Employee Compliance – Legal Matters – Includes assuring the compliance with all corresponding laws regulating the recruitment, management, and termination of employees.
10. Employee Relations – Labour Issues – Includes handling legal concerns of employees' contracts and negotiations, along with the establishing negotiation team and strategies (Jones & Walters, 1994).
11. **Information Systems and Technology:** Includes providing human resources focused technology to enhance the quality of services when recruiting, while maintaining compliance and empowering professional development and retention.
12. **Employee Leave and Holiday**: Includes non-work activities of employees by allowing paid or unpaid leisure, whether required by policy or designed by the employer.
13. **Payroll:** Includes the determination of compensation.
14. **Retirement:** Includes providing plans for retirement comprising analysis of plans and implementation procedures (Jones & Walters, 1994).
15. **Training and Development:** Includes providing training and development programmes that meet the employees' needs.

In an attempt to identify the best practices, human resource professionals, and experts have spent a lot of efforts for validating human resource strategies and policies. The best practices of human resource management could be defined as those functions that evidently promote human and financial performances (Hafford & Moore, 2005). The purposeful cycle of development of information technology as innovation implementation can lead to the effective application of information technology (Bilevičienė, Bilevičiūtė & Paražinskaitė, 2015).

CONCLUSION

Many researchers have suggested that creativity makes an important contribution to organizational effectiveness for the long-term survival of organizations, because it enables organizations to remain competitive in a rapidly changing environment and achieve a competitive advantage. Thus, encouraging and fostering creativity is a strategic choice of every successful organization. In this competitive business world every organization whether academia or industry is in the need of develops their operational activities in order to sustain in the market. There are emerging trends or innovations to be followed in HRM to improve their efficiency in providing innovational activities in the organization. So HR department increasingly adopt open innovation models and engage with external knowledge sources and they want to bring new groups into the innovation process. This leads for dedicated training of employees, new performance indicators, new rewards, new ways of communicating with and between employees etc., The HRM innovations followed in the organization will have positive influence on the innovation performance among the employees and brings great impact in development. By 2050 the business world will have drastic changes in its functioning. The changes in the technology, global economy, increasing ability scrutiny, threatening talent crisis and the mental illness of the employees are drastically affecting the workplace. These changes will have great impact on the business environment. The best organizations with the sustained innovation derive success towards the competition.

REFERENCES

Ambardar, A., & Singh, A. (2017). Quality of Work Life Practices in Indian Hotel Industry. *International Journal of Hospitality and Tourism Systems*, *10*(1), 22.

Analoui, F. (2007). *Strategic Human Resource Management, Thomson Learning*. Ashgate.

Andresen, J., Baldwin, A., Betts, M., Carter, C., Hamilton, A., Stokes, E., & Thorpe, T. (2002). A framework for measuring IT innovation benefits. *Journal of Information Technology in Construction*, 5(4), 57–72.

Araújo, L. C. G. d., & Garcia, A. A. (2009). Gestão de pessoas: estratégias e integração organizacional. Academic Press.

Bilevičienė, T., Bilevičiūtė, E., & Paražinskaitė, G. (2015). Innovative Trends in Human Resources Management. *Economia e Sociologia*, 8(4), 94–109. doi:10.14254/2071-789X.2015/8-4/7

Chan Kim, W., & Mauborgne, R. (2005). *Blue Ocean Strategy: How to create uncontested market space and make the competition irrelevant.* Harvard Business Review Press.

Damanpour, F., & Gopalakrishnan, S. (1998). Theories of organizational structure and innovation adoption: The role of environmental change. *Journal of Engineering and Technology Management*, 15(1), 1–24. doi:10.1016/S0923-4748(97)00029-5

EBOLI. (2013). Educação Corporativa Desenvolvendo a Excelência Profissional e Organizacional. *AGANP Proceedings, Goiânia.*

García-Carbonell, N., Martín-Alcázar, F., & Sánchez-Gardey, G. (2015). Determinants of top management's capability to identify core employees. *BRQ Business Research Quarterly*, 18(2), 69–80. doi:10.1016/j.brq.2014.07.002

Geer, D. E., Jr., Tumblin, H. R., & Solomon, E. M. (2001). Enabling business transactions in computer networks. Google Patents.

Hafford, J. C., & Moore, J. E. (2005). *Sourcing Best Practices in Human Resources.* SHRM white paper.

Hamel, G. (2008). The future of management. *Human Resource Management International Digest*, 16(6), hrmid.2008.04416fae.001. doi:10.1108/hrmid.2008.04416fae.001

Hartmann, A. (2006). The role of organizational culture in motivating innovative behaviour in construction firms. *Construction Innovation*, 6(3), 159–172. doi:10.1108/14714170610710712

Joseph, R. C., & Ezzedeen, S. R. (2009). E-government and e-HRM in the public sector. In Encyclopedia of Human Resources Information Systems: Challenges in e-HRM (pp. 272-277). IGI Global. doi:10.4018/978-1-59904-883-3.ch041

Jussani, A. C., Krakauer, P. V. C., & Polo, E. F. (2010). Reflexões sobre a estratégia do oceano azul: Uma comparação com as estratégias de Ansoff, Porter e Hax & Wilde. *Future Studies Research Journal: Trends and Strategies, 2*(2), 17–37.

Katou, A. A. (2015). The mediating effects of psychological contracts on the relationship between human resource management systems and organisational performance. *International Journal of Manpower, 36*(7), 1012–1033. doi:10.1108/IJM-10-2013-0238

Kavanagh, M. J., & Johnson, R. D. (2017). *Human resource information systems: Basics, applications, and future directions.* Sage Publications.

Khazanchi, S., Lewis, M. W., & Boyer, K. K. (2007). Innovation-supportive culture: The impact of organizational values on process innovation. *Journal of Operations Management, 25*(4), 871–884. doi:10.1016/j.jom.2006.08.003

Laursen, K., & Foss, N. J. (2003). New human resource management practices, complementarities and the impact on innovation performance. *Cambridge Journal of Economics, 27*(2), 243–263. doi:10.1093/cje/27.2.243

Nacinovic, I., Galetic, L., & Cavlek, N. (2009). Corporate culture and innovation: Implications for reward systems. *World Academy of Science, Engineering and Technology, 53,* 397–402.

Navin, A. H., Navimipour, N. J., Rahmani, A. M., & Hosseinzadeh, M. (2014). Expert grid: New type of grid to manage the human resources and study the effectiveness of its task scheduler. *Arabian Journal for Science and Engineering, 39*(8), 6175–6188. doi:10.100713369-014-1256-7

Noe, R. A., Hollenbeck, J. R., Gerhart, B., & Wright, P. M. (2017). *Human resource management: Gaining a competitive advantage.* McGraw-Hill Education New York.

Panigrahy, P., & Pradhan, K. (2015). *Creativity and innovation: Exploring the role of HR practices at workplace.* Paper presented at the Presentation of Paper at National Conference organized by Ravenshaw B-School, Cuttack.

Papageorgiou, D. (2018). Transforming the HR function through robotic process automation. *Benefits Quarterly, 34*(2), 27–30.

Parasuraman, R., & Riley, V. (1997). Humans and automation: Use, misuse, disuse, abuse. *Human Factors, 39*(2), 230–253. doi:10.1518/001872097778543886

Rothwell, R. (1992). Successful industrial innovation: Critical success factors for the 1990''s. *R & D Management, 22*(3), 221–239. doi:10.1111/j.1467-9310.1992.tb00812.x

Rousseau, D. M., & Wade-Benzoni, K. A. (1994). Linking strategy and human resource practices: How employee and customer contacts are created. *Human Resource Management, 33*(3), 436–489. doi:10.1002/hrm.3930330312

Ruël, H., Bondarouk, T., & Looise, J. K. (2004). E-HRM: Innovation or irritation. An explorative empirical study in five large companies on web-based HRM. *Management Review*, 364–380.

Schalk, T. B., Stovall, J. L., & Brooks, W. P. (2011). *Multi-modal automation for human interactive skill assessment*. Google Patents.

Schraft, R.-D., & Schlaich, G. (1988). A survey of the assembly of wire harnesses in industry. *Assembly Automation, 8*(1), 29–32. doi:10.1108/eb004230

Schuler, R. S. (1992). Strategic human resources management: Linking the people with the strategic needs of the business. *Organizational Dynamics, 21*(1), 18–32. doi:10.1016/0090-2616(92)90083-Y

Serafim, L. (2011). *O Poder da Inovação: a experiência da 3m e de outras empresas Inovadoras*. Saraiva.

Sheridan, T. B. (1992). *Telerobotics, automation, and human supervisory control*. MIT press.

Thom, N., Ritz, A., & Masiulis, K. (2004). Viešoji vadyba: inovaciniai viešoji sektoriaus valdymo metmenys. Verlag der Rechtswissenschaftlichen Universität Litauen.

Tung, R. L., Paik, Y., & Bae, J. (2011, January). Korean HRM in the global context. *International Journal of Human Resource Management, 22*(2), 481–482. doi:10.1 080/09585192.2011.543315

Vveinhardt, J., & Andriukaitiene, R. (2016). Model of establishment of the level of management culture for managerial decision making with the aim of implementing corporate social responsibility. *Transformations in Business & Economics*, 15.

Walke, S. G. (2013). *Critical study of agritourism industry in Maharashtra*. Academic Press.

Wolfe, R., Wright, P. M., & And Dennis, L. S. (2006, Spring). Radical Hrm Innovation And Competitive Advantage: The Moneyball Story. *Human Resource Management, 45*(1), 111–145. doi:10.1002/hrm.20100

Woodman, R. W., Sawyer, J. E., & Griffin, R. W. (1993). Toward a theory of organizational creativity. *Academy of Management Review*, *18*(2), 293–321. doi:10.5465/amr.1993.3997517

Ying, S. (2006). *Creating Supportive Environment for Innovation: A Conceptual Model Study*. Academic Press.

Chapter 7

Investigating the Effect of Information Technology and Key Determinants on Knowledge Sharing and Arriving at Employee Creativity

Mahendra Kumar Sharma
Indore Institute of Management and Research, Indore, India

Dilip Kumar
ICFAI University Jharkhand, India

ABSTRACT

The chapter aims to investigate the influence of information technology, trust, rewards, leadership, and organizational culture on the knowledge sharing behavior of the employees that ultimately drives employee creativity. Drawing from the literature on employee creativity, knowledge sharing, and its influencing variables, this paper proposed a model comprising all such prominent variables and tested it quantitatively. For this purpose, 405 questionnaires were collected at Indore, India, and structural equation modeling was used to test the hypotheses. The findings show that organizational culture followed by leadership was the prominent factor affecting the knowledge sharing behavior of employees. Information technology, trust, and rewards followed next, respectively. Employee creativity was found to be significantly affected by knowledge sharing behavior. The study augments the research on employee creativity and knowledge sharing.

DOI: 10.4018/978-1-7998-4180-7.ch007

INTRODUCTION

Failure to learn and subsequently innovate may turn an organization into sub-optimal or even make it dysfunctional (DiBella, 1995), whereas a learning organization is expected to move ahead with better performance in the market.

In the present competitive era, despite having financial soundness and technological advancements, human capital is considered to be the most crucial competency an organization can hold who in turn substantially influence the usage of other resources of the organization. The employees of an organization hold certain education about their field and over a period of time they gain expertise in the field. At times, there are situations where working can be made easy and complex situations and problems handled in a manner not mentioned in any book but only learned by experience of self and others. Competent employees are seen engaging themselves in searching, acquiring, sharing and exploiting their knowledge to help the organization flourish (Henri, 2016). This knowledge base is nothing less than a mine of valuable asset that the employee can explore and pass on to others in order to make others' job easy as well.

Knowledge can be referred to as an individual's comprehension of an area of interest which has been gained by experience and study (Awad & Ghaziri, 2004). Knowledge is a very essential aspect intellectual asset of an organization (Grant, 1996a, 1996b) that powers its competitive advantage in the ever increasing market competition (Argote*et al.*, 2003)and is rightly considered to be a crucial strategic resource held by an organization (Chen *et al.*, 2016)..

Knowledge sharing can be defined as the process in which mutual exchange of knowledge among individuals, teams, departments and organizations take place giving rise to new knowledge (Ipe, 2003; van den Hooff, Elving, Meeuwsen, &Dumoulin, 2003). Knowledge sharing can also be defined as "processes that comprise exchange of knowledge between individuals and groups" (Yu *et al.*, 2010). It can further be comprehended as the provision of job, information and required know-how to assist others and also to cooperate with such others when it comes to solving of problems, developing new ideas, or implementing policies or procedures (Wang & Noe, 2010).

Knowledge sharing is considered as a key factor governing the flow of information and experience in the organization. On one side is the supply side of the process where employees in the organization are motivated to share with others their knowledge in order to benefit other employees and the organization as well. The other side of this is the demand side which is concerned with the behavior of knowledge sharing among the employees and the knowledge acquisition in order to improve the knowledge base of the organization (Mansingh*et al.,* 2009).

Proper knowledge sharing activities help in the development of competencies and the knowledge itself among the employees, adding to the advantage of the organization in the competitive world (Fierro*et al.*, 2011; Suresh, 2012).

Knowledge sharing across the organization is a must (Wang and Noe, 2010), that helps in better performance of an organization (Rutten, Blaas-Franken, & Martin, 2016), facilitates the overall effectiveness of the organization (Renzl, 2008), is crucial in determining and enhancing an organization's capacity to innovate Zhang *et al.*, 2018; Le and Lei, 2018; Bontis*et al.*, 2009; Svetlik), assists in idea generation activities (Henri, 2016), aids persistent success and innovation (Drucker, 1999; Nonaka& Takeuchi, 1995), increases productivity and performance (Cummings,2004; Lin, 2007; Mesmer-Magnus &DeChurch, 2009), increases employee performance and organizational teamwork (Cabrera and Cabrera, 2005; Cummings, 2004; Hansen, 2002; Mesmer-Magnus &DeChurch, 2009), gains gratitude by colleagues (Martinez, 2015); helps develop competitive advantages for the organization (Lee &Lan, 2011; Liu & Deng, 2015; Loureiro*et al.*, 2015; Marqués&Simón, 2006; Rahimli, 2012) and achieve the stated organizational goals more efficiently (Le and Lei, 2017).

Improper management and hoarding of knowledge result in the corroding of knowledge. Knowledge sharing should be understood as a building block in the organization's success (Witherspoon, Bergner, Cockrell, & Stone, 2013). However, a fundamental issue is the individual's reluctance to engage in knowledge sharing behavior out of the fear of losing knowledge ownership and associated power (Alsharo*et al.*, 2017).

Creativity can be understood as birth of new and useful ideas by one person or a group working collectively (Zhou and George, 2001), that is worth supporting the organization in the market (Gu*et al.*, 2015). This creativity is an outcome of distinguishing features of individuals (Chang *et al.*, 2014) and has attracted its share of attention from practitioners and scholars (Chang*et al.*, 2014). The knowledge sharing practices in an organization have a profound influence on the employees' creativity and the resulting effectiveness in operations in leading the organization (Inkinen, 2016).

Continuous innovation and addition to the existing knowledge has been well identified by practitioners and academic scholars as a vital ingredient if the organization wishes to make a mark in the competitive market (DiPietro&Anoruo, 2006; Mumford, 2000; Weiner, 2000). Factors like organizational culture (Hoof&Huysman, 2009; Huysman&Wulf, 2006), management support/leadership (Connelly &Kevin Kelloway, 2003), state of technology (Carlson & Davis,1998;Hooff&Huysman, 2009) are among the key factors influencing the knowledge sharing behavior of an individual in the organization.

On one hand knowledge sharing behavior is believed to enhance when clubbed with rewards and incentives (Hislop, 2009), whereas at times it has been observed that such incentives scarcely or don't affect the knowledge sharing behavior of the individuals. Many organizational factors like organizational culture and leadership are among the prime influencers of knowledge sharing behavior. At times, there are conflicting results regarding certain factors in consideration as well. The aim of this paper is to explore the key antecedents of knowledge sharing behaviour that are relevant in the modern era for facilitating the passing of knowledge in the organizational setup and also to assess the impact of each antecedent on knowledge sharing behaviour. Furthermore, the paper makes an attempt to uncover the extent of influence that knowledge sharing has on employee creativity.

LITERATURE REVIEW AND HYPOTHESES DEVELOPMENT

Organizational Culture

Organizational culture can be understood as the thinking orientation, shared beliefs, principles and shared values that people in an organization hold in terms of their behavior towards other people in the organization (Wei, The&Asmawi, 2012). It is a phenomenon on group-level whose perception varies from organization to organization (Shin *et al.*, 2016) perhaps as a result of the varied dimensions it demonstrates (Ellen and Nico, 2002; Hamza*et al.*, 2011; Hoskins, 2014; Porter *et al.*, 2016) and is also guided by the perception of its definition (Schein, 2010). The values may welcome or deny fresh ideas (Chang and Nadine, 2014) and are profoundly placed in the employees' mind of an organization. (Martins and Terblanche, 2003).

Organizational culture plays a vital role and acts as a key factor that helps promote knowledge sharing delightedly within an organization (Alam, *et al.*, 2009; Hooff&Huysman, 2009; Huysman&Wulf, 2006) and also across organizations. A positive and enabling organizational culture is essential to facilitate the knowledge sharing behavior in an organization (Connelly &Kevin Kelloway, 2002; Govindarajan& Gupta, 2000; Gummer, 1998).

A positive organizational culture acts as an aid for the employees to engage in knowledge sharing behavior (Anantatmula, 2007) whereas an unfavourable organizational culture tends to act as a hindrance in the knowledge sharing process (David & Fahey, 2000; McDermott &O'dell, 2001). The above discussion highlights clearly the role of conducive organizational culture in knowledge sharing behavior among the employees. Thus, the following hypothesis can be framed:

H1: Organizational culture positively impacts the knowledge sharing behavior of employees.

TRUST

Trust can be understood as the expectation an individual develops as a result of honest, regular and cooperative behavior in the society/community based on some common norms and members of the particular society/community (McElroy *et al.*, 2006).

Trust is considered a very crucial element that acts as motivators for successful knowledge sharing (Ford, 2003; Rolland &Chauvel, 2000) and assists in the collaboration strengthening among the employees(Politis, 2003). Trust is regarded as a driving force of knowledge sharing (Rutten*et al.* 2016). Trust can be understood as a medium by which knowledge can be exchanged easily and emphasizing on trust building in the organization can lead to an increased knowledge sharing in the organization (Huang *et al.*, 2008).

Individuals having higher level of trust are more inclined towards taking risk in knowledge sharing (Nahapiet&Ghoshal, 1998). Trust is believed to lower any such possible detrimental consequence in the apparent cost of knowledge sharing (Wei, The&Asmawi, 2012). Trust is considered to be the most efficient and least expensive tool for reinforcing employees to engage in knowledge sharing behavior (Alam, *et. al.*, 2009). Knowledge sharing is directly and positively affected by trust (Holste and Fields, 2010; Jain *et al.*, 2015; Koohang*et al.,* 2017; Le and lei, 2018). An organization can only enhance the activities of sharing knowledge among employees to foster its innovation capability if interpersonal trust is built (Donate and Guadamillas, 2011; Alsharo*et al.*, 2017; Zhang *et al.*, 2018; Guadamillas, 2011; Le and Lei, 2018; Yang *et al.*, 2018). The following hypothesis is proposed:

H2: An employee's trust level positively influences his knowledge sharing behavior in the organization

Leadership

A leader is seen as a role model exhibiting ethical and principal oriented acts (Mayer *et al.*, 2009; Schaubroeck*et al.*, 2012) and is expected to facilitate knowledge sharing among the employees (Kerr & Clegg, 2007). Such individuals possess knowledge and understanding of making use of their skills (Walumbwa*et al.*, 2008) and can estimate the influence of their judgments on the followers (Luthans and Avolio, 2003; Gardner *et al.*, 2011).

Leadership has attracted its fair share of attention from the practitioners and academicians alike when it comes to exploring its importance in the knowledge sharing behavior of employees in the organization (Brown and Treviño 2006; Bryman, 2007; Hofmeyer*et al.* 2015; Rahaman*et al.*, 2019). Leadership can be understood to be concerned with the process of influencing and including others towards attaining desired goals (Dorfman and House, 2004; Javidan and Carl, 2005).

Leadership is considered to be of prime importance while understanding the knowledge sharing behaviour (Leithwood*et al.*, 1999). A noteworthy association exists between leadership and the knowledge sharing behavior (Kark*et al.*, 2003; Srivastava*et al.*, 2006; Lin, 2007a, 2007b; Birasnav*et al.*, 2011). The interaction between the employees and their superiors has the tendency to influence greatly the knowledge sharing among them (Wei, The, & Asmawi, 2012). Effective leadership has the capacity to strongly influence knowledge sharing behavior of others. Following the discussion the hypothesis is framed as below:

H3: Leadership positively influences the knowledge sharing behavior of employees

Information Technology

Knowledge sharing can never take place in isolation and in today's world connectivity is all the more necessary at individual level to exchange information and knowledge. Connectivity can be understood as the ability of the members of a system to contact each other (van den Hooff, Elving, Meeuwsen, &Dumoulin, 2003). Information technology comes in handy as a medium to provide better and faster mode of connectivity (Riege, 2005). Information technology infrastructure can be understood as the compilation of the essential technology tools needed to support the knowledge sharing efforts. Technology facilitates the employees to access other employees with essential expertise to attend and solve the problems at hand. Information technology provides for all the essential collaboration, communication and networking capacity needed for the proper knowledge creation and its transfer (Al-Ammary, Fung, &Goulding, 2008).

The availability and usability of technology have ample influence on the knowledge sharing in an organization (Anantatmula, 2007; Smith, 2001). Technology can be viewed as a facilitator for knowledge sharing in the organizational setup. The development in information technology in the recent years have made it easier for the organization to receive and pass the required information to the various stakeholders and also assisted in information passing within the organization (Tseng, 2008). The resulting hypothesis is:

H4: The availability and access to information technology positively impacts the knowledge sharing behavior of employees

Reward

Reward system of an organization can be said to determine as to how knowledge is accessible and flows in an organization (Leonard-Barton, 1998). Rewards are considered to be helpful in recognizing the excellent knowledge sharing abilities of the employees. As a result, fair and objective reward system should be in place to motivate the employees (Grayson & O'Dell, 1998).

Rewards are considered to be among the most effective methods of motivating employees to engage in knowledge sharing behaviour with other employees (Alam, *et al.*, 2009). Performance-based system contributes significantly to the knowledge sharing behavior of employees within the organization (Kim & Lee, 2006). Rewards such as salary hike, promotions and bonus are positively associated with the knowledge sharing behavior of employees (Kankanhalli*et al.*, 2005).Rewards have a positive impact on the knowledge sharing in organizations (Bartol&Srivastava, 2002). However, withdrawal of rewards is followed by subsequent decrease in the knowledge sharing.The following hypothesis is framed:

H5: Perception of fair reward system in the organization positively impacts the knowledge sharing behavior of employees

Employee Creativity

Organizations are understood to achieve success comfortably in a situation when they identify and accord individual creativity and take steps to nurture and promote employee creativity (Williamson, 2001). Creativity, an individual-level activity (Hennessey and Amabile, 2010), is thought to be the foundation of innovation (Dewett&Gruys 2007). If properly facilitated, employee creativity can act as a building block for organizational change, innovation and competitiveness (Mumford, 2000; Williamson, 2001; Zhou & George 2001; DiPietro&Anoruo, 2006) as an individual is considered to be at the center of novel idea generation in an organization (Gilad, 1984; Whiting, 1988; Mumford, 2000).

Talking of creativity of an employee in the organizational setup, need of various resources is highlighted in the literature. Time, teamwork effort and knowledge resources being among them. Knowledge here is identified as a crucial and essential resource that in fact facilitates and stimulates employee creativity (Gardner &Laskin, 2011; Akturan and Çekmecelio_glu (2016). It also helps in enhancing performance, current knowledge abilities and innovation (Yeh*et al.*, 2012; Lee and Hong, 2014).

Knowledge sharing acts as a vital link for the passage of information and knowledge; which are considered to be of utmost importance for the employee creativity (Shalley*et al.*, 2004).Knowledge sharing is closely related to the creativity of employees they apply in their jobs for the organization where new knowledge development and innovative ideas are sought (Cabrera & Cabrera, 2005; Shazi*et al.*, 2015; Wang *et al.*, 2017; Hui*et al.,* 2018; Lei *et al.,* 2018; Yang*et al.,* 2018). Knowledge sharing can help people gain high quality information and knowledge and by combining these with their own knowledge, employees can finally come forward with more refined knowledge and creative ideas (Amin *et al.*, 2011).

H6: Knowledge sharing positively influences the employee creativity.

Figure 1. Conceptual model showing various relationships between the constructs

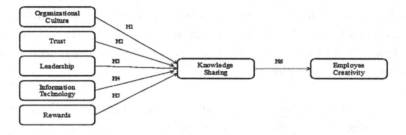

RESEARCH METHODOLOGY

Questionnaire

The questionnaire used for the study was divided in two sections. The first section dealt with information about the socio-demographic features of the respondents, like – age, sex, marital status, income, seniority, department, etc. The next section of the questionnaire included statements regarding various constructs included in the study. Open-ended multiple statements were used for each factor to collect response from the respondents. The response was collected with the help of a five-point Likert scale ranging from 'strongly disagree' (5) to 'strongly agree' (1).SPSS and AMOS software were used to assess the proposed theoretical framework.

Measures

After extensive literature review various sources were inculcated to draft the final questionnaire. The items used to measure the constructs in the study were adapted to suit the context from previous studies the details of which are mentioned in **Table 1.** The questionnaire was initially reviewed by subject experts and the potential respondents. With minor modifications, the questionnaire was put to a pilot study with a small sample size of 50 respondents to assess parameters like reliability of constructs, factor analysis. The factor loadings of the items were well above the .60 mark and the Coefficient Alpha for the constructs met the minimum acceptable value of .70 making the questionnaire fit for further use **Table 6.**

Data Collection

The study used non-probability convenience sampling method to collect the data. The questionnaire was administered to the respondents working at lower, middle and top management in various companies in the cities of Indore and Bhopal, India. In total, 34 companies were approached for the study(each havingminimum 300 employees). In a total of 417 filled questionnaires over a period of three weeks, a total of 405 valid and complete questionnaires were identified. The respondentsprofiles are mentioned in **Table 2.**

DATA ANALYSIS

Structural Equation Modeling was used for model validation. SEM enables the researcher to test causal relationships between various constructs with multiple measures (Joreskog and Sorbom, 1996). The study followed the sequential six-step approach for the same (Hair et al., 2014).

Descriptive Statistics

The reliability of all the constructs was assessed using the Cronbach's Alpha parameters. Organizational culture with 0.874, leadership with 0.875, information technology with 0.879, trust with 0.888 and rewards with 0.889, knowledge sharing behavior with 0.890 and employee creativity with 0.876 were found to be fit enough (being above the 0.7 mark) for further data analysis.

Exploratory Factor Analysis

Exploratory Factor Analysis using principal component analysis with varimax rotation method was applied. Items with factor loading above 0.50 were considered to determine items clubbed into a single factor. Further, measure of sampling adequacy through KMO comes out to be .927 and Bartlett's test of Sphericity indicate the Chi-square of 8962.128 with df= 406, p= .000<0.05]. These values indicate the appropriateness to proceed with factor analysis.

Table 1. References of construct related items used in the questionnaire

Sl. No.	Construct	References
1.	**Organizational Culture**	Singh andBhandarkar (2011) Tsui*et al.* (2006),
2.	**Trust**	Chiu *et al.* (2006), Von-Krogh (1998)
3.	**Leadership**	Arnold *et al.* (2000) Raquib*et al.* (2010),
4.	**Information Technology**	Kim and Lee (2006) Menolli*et al.* (2015) Yuan *et al.* (2013
5.	**Reward**	Srivastava*et al.* (2006) Sveiby (2001)
6.	**Employee creativity**	Zhou andGeorge (2001). Baer andOldham (2006) Gong *et al.* (2009) Zhou and George (2001).
7.	**Knowledge Sharing**	Alam*et al.* (2009) Bowen (2010) Chai *et al.* (2011) Fathi*et al.* (2011)

Table 2. Cumulative respondents profile participating in the survey

Demographic Features	No. of Respondents	Percent
Position		
Top level manger	32	8%
Middle level manager	327	81%
Lower level manager	46	11%
Department		
Marketing	245	60%
Operations	37	9%
Administrations	68	17%
Others	55	12%
Years served in the company		
0-1 year	89	22%
1-3 years	238	59%
3-5 years	48	12%
5-8 years	16	4%
8 or more years	14	3%

Table 3. Reliability

Case Processing Summary			
		N	%
Cases	Valid	385	100.0
	Excluded[a]	0	.0
	Total	385	100.0

a. Listwise deletion based on all variables in the procedure.

Table 4.

Reliability Statistics	
Cronbach's Alpha	Number of Items
.925	29

Measurement Model Through Confirmatory Factor Analysis

In the above table, reliabilities of all the constructs range from 0.890 to 0.804 and this meets the minimum required benchmark of 0.7 implying a good reliability (Hair *et al.*, 2008). The construct validity is concerned with the facet if a measurement scale is a suitable operational description of a particular construct (Flynn, Sakakibara, Schroeder, Bates, & Flynn, 1990).

Structural Equation Model

This is an analytical approach to study the linkages between latent variables or unobserved variables and manifest or observed variables which constitute the unobserved variables. In SEM, there is an inimitable feature of being able to contain variables that are not directly assessable and thus called unobserved or latent construct.

Table 5. KMO and Bartlett's Test

KMO and Bartlett's Test		
Kaiser-Meyer-Olkin Measure of Sampling Adequacy.		.927
Bartlett's Test of Sphericity	Approx. Chi-Square	8962.128
	Df	406
	Sig.	.000

Table 6. Summarized Exploratory Factor Analysis Results

Factor	Items	Factor Loading	Reliability	Eigenvalue	Variance Explained	Communality
Organizational Culture (OC)			0.874	9.716	33.502	
	OC_1	0.751				0.745
	OC_2	0.775				0.686
	OC_3	0.804				0.731
	OC_4	0.789				0.758
Leadership (LES)			0.875	2.743	9.457	
	LE_1	0.819				0.743
	LE_2	0.797				0.696
	LE_3	0.819				0.732
	LE_4	0.826				0..759
Information Technology (IT)			0.879	2.284	7.874	
	IT_1	0.841				0.747
	IT_2	0.848				0.742
	IT_3	0.820				0.726
	IT_4	0.830				0.738
Trust (TR)			0.888	2.163	7.457	
	TR_1	0.785				0.729
	TR_2	0.754				0.696
	TR_3	0.767				0.697
	TR_4	0.774				0.684
	TR_5	0.776				0.725
Rewards (RE)			0.889	1.706	5.883	
	RE_1	0.828				0.740
	RE_2	0.817				0.735
	RE_3	0.835				0.751
	RE_4	0.858				0.790
Knowledge Sharing Behaviour (KS)			0.890	1.507	5.196	
	KS_1	0.775				0.795
	KS_2	0.726				0.729
	KS_3	0.714				0.726
	KS_4	0.735				0.771
Employee Creativity (EC)			0.876	1.153	3.975	
	EC_1	0.796				0.732
	EC_2	0.782				0.710
	EC_3	0.775				0.725
	EC_4	0.793				0.760

Table 7. Construct Reliability, Convergent validity and Discriminant Validity

	CR	AVE	MSV	MaxR(H)	EC	OC	IT	LES	RE	KS	TR
EC	0.877	0.640	0.352	0.878	**0.800**						
OC	0.875	0.636	0.353	0.878	0.508	**0.797**					
IT	0.879	0.645	0.194	0.879	0.287	0.300	**0.803**				
LES	0.876	0.639	0.270	0.878	0.395	0.317	0.163	**0.799**			
RE	0.890	0.670	0.195	0.892	0.364	0.355	0.243	0.258	**0.818**		
KS	0.892	0.673	0.353	0.894	0.593	0.594	0.441	0.520	0.442	**0.821**	
TR	0.889	0.616	0.326	0.891	0.511	0.571	0.194	0.437	0.353	0.555	**0.785**

In order to formulate hypothesis, structural equation modeling was applied to check if the data supported the proposed model.

Table 8. Estimates (SRWs) of Items

Sl. No	Label	Construct	Estimates	Sl. No.	Label	Construct	Estimates
1	OC_4 <---	F1	.832	15	TR_3 <---	F4	.762
2	OC _3 <---	F1	.772	16	TR_2 <---	F4	.787
3	OC _2 <---	F1	.759	17	TR_1 <---	F4	.742
4	OC _1 <---	F1	.824	18	RE_4 <---	F5	.857
5	LES_4 <---	F2	.820	19	RE_3<---	F5	.807
6	LES _3 <---	F2	.761	20	RE_2 <---	F5	.802
7	LES _2 <---	F2	.790	21	RE_1 <---	F5	.806
8	LES _1 <---	F2	.825	22	KS_4 <---	F6	.839
9	IT_4 <---	F3	.808	23	KS_3 <---	F6	.808
10	IT_3 <---	F3	.789	24	KS_2 <---	F6	.790
11	IT_2 <---	F3	.798	25	KS_1 <---	F6	.844
12	IT_1 <---	F3	.816	26	EC_4 <---	F7	.829
13	TR_5 <---	F4	.816	27	EC_3 <---	F7	.794
14	TR_4<---	F4	.815	28	EC_2 <---	F7	.778
				29	EC_1 <---	F8	.798

Assessment of Model Fit Indices for the Structural Model

The figure revealed that the structural equation model comprising all the variables and their relationships. The measurement model reveals a Normed Chi-square of 3.462, GFI= 0.826, AGFI= 0.792, RMSEA= 0.066, NFI= 0.912, TLI= 0.941, CFI= 0.947, IFI= 0.947, PCFI= 0.865 and PNFI= 0.853. The various model fit indices meet the least measures of a good model fit.

Table 9. Model Fit indices for structural model for Customer Buying Decision

Type of Measure	Model Fit Indices	Model Value
Absolute Fit Measure		
	CMIN	1295.042
	DF	374
	CMIN/DF	3.462
	GFI	0.826
	AGFI	0.792
	RMSEA	0.086
Incremental Fit Measure		
	NFI	0.912
	TLI	0.941
	CFI	0.947
	IFI	0.947
Parsimony-Adjusted Measures		
	PCFI	0.865
	PNFI	0.853

Hypothesis Testing Through Path Analysis

H1: Organizational culture impacts the knowledge sharing behavior of employees.

In Table no. 10 the p-value stands at 0.000, which is less than the significance level of 5%, so we accept the alternate hypothesis. In other words, there is significant relationship between the organization culture and knowledge sharing behavior of employees.

H2: An employee's trust level influences his knowledge sharing behavior in the organization

In Table no. 10 the p-value stands at 0.000, which is less than the significance level of 5%, so we accept the alternate hypothesis. In other words, there is significant relationship between the trust level of employees and knowledge sharing behavior of employees.

H3: Leadership influences the knowledge sharing behavior of employees

In Table no. 10 the p-value stands at 0.000, which is less than the significance level of 5%, so we accept the alternate hypothesis. In other words, there is significant relationship between the leadership in the organization and knowledge sharing behavior of employees.

H4: The availability and access to information technology impacts the knowledge sharing behavior of employees

In Table no. 10 the p-value stands at 0.000, which is less than the significance level of 5%, so we accept the alternate hypothesis. In other words, there is significant relationship between the availability and access of information technology and knowledge sharing behavior of employees.

H5: Perception of fair reward system in the organization impacts the knowledge sharing behavior of employees

In Table no. 10 the p-value stands at 0.000, which is less than the significance level of 5%, so we accept the alternate hypothesis. In other words, there is significant relationship between the perception of fair reward system in the organization and knowledge sharing behavior of employees.

H6: Knowledge sharing influences the employee creativity.

In Table no. 10 the p-value stands at 0.000, which is less than the significance level of 5%, so we accept the alternate hypothesis. In other words, there is significant relationship between the knowledge sharing behavior of employees and employees creativity.

Figure 2. Path Analysis Model

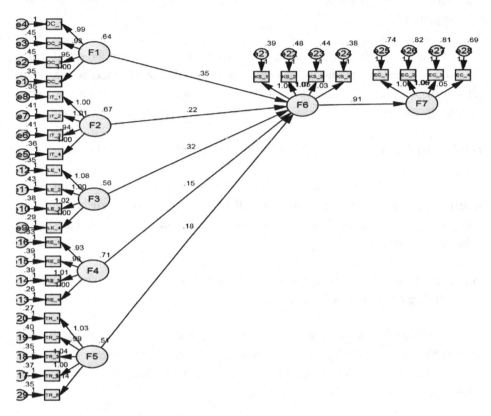

5. CONCLUSION

Knowledge is a very important determinant for the organization to maintain and build competency in today's competitive world. An organization that maintains a

Table 10. Path Estimates (SRWs), standard error, critical region and p value

Path Direction	Estimate	SE	Critical Region	P Value	Decision
OC→KS	0.348	0.029	11.861	0.000	Not Accepted*
LE→ KS	0.225	0.023	9.770	0.000	Not Accepted*
IT→ KS	0.323	0.029	11.208	0.000	Not Accepted*
TR→ KS	0.145	0.019	7.568	0.000	Not Accepted*
RE→ KS	0.183	0.024	7.749	0.000	Not Accepted*
KS→EC	0.909	0.106	8.583	0.000	Not Accepted*

* The p value lies below 0.05 at 95% confidence level which suggests that the hypothesis cannot be accepted.

healthy atmosphere where the employees are motivated to share knowledge surely creates opportunity to excel and have a smooth functioning.

In all the antecedents of knowledge sharing studied in the paper, organizational culture was found to be the most important influencer. This shows how important an organization's employee policies, organizational structure and working style deeply impacts the overall understanding of the employee to engage in knowledge sharing or not. It was followed by leadership an employee faces in the organization that most prominently influences the knowledge sharing of employees. In the order of decreasing order the other constructs influencing knowledge sharing behavior are information technology, trust, and rewards. The organization should develop ample technological support to enable employees share information with each other efficiently and effectively. Fair and objective reward system should be in place to motivate the employees to engage in knowledge sharing behavior.

Positive knowledge sharing behavior helps employees to work on increasing and polishing their creativity at job that again benefits the organization. Creativity being of utmost importance to help organization innovate and make its working more efficient making the organizational activities more profitable

LIMITATIONS AND FUTURE RESEARCH

Geographical coverage is a prominent limitation of the present study. Increased number of sample size would have been an advantage. Additionally, researcher made attempt to inculcate ample and most relevant constructs in the study surrounding knowledge sharing and employee creativity but more constructs can be explored on the same platform. Although the present study limits itself to knowledge sharing leading to employee creativity, a more comprehensive approach can be used in the future studies to focus on knowledge management system as a whole in an organization.

Future studies can also undertake a comparative analysis about knowledge sharing orientation of employees at the time of joining the organization and then after certain years of service. This may also serve as an appraisal tool for the policies of the organization itself.

REFERENCES

Al-Ammary, J. H., Fung, C. C., & Goulding, P. (2005). Alignment of knowledge and IS/IT strategies: A case for the banking sector in the Gulf Cooperation Countries (GCC). *International Conference on Knowledge Management (ICKM 2005)*.

Alam, S., Abdullah, Z., Ishak, N., & Zain, Z. (2009, April). Assessing Knowledge Sharing Behaviour among Employees in SMEs: An Empirical Study. *International Business Research*, *2*(2), 2. doi:10.5539/ibr.v2n2p115

Amin, Basri, Hassan, & Rehman. (2011). *Occupational stress, knowledge sharing and GSD communication barriers as predictors of software engineer's creativity.* Academic Press.

Anantatmula, V. S. (2007). Linking KM effectiveness attributes to organizational performance. *Vine*, *37*(2), 133–149. doi:10.1108/03055720710759928

Awad, E. M., & Ghaziril, H. M. (2004). *Knowledge Management.* Pearson Education.

Bartol, K. M., & Srivastava, A. (2002). Encouraging knowledge sharing: The role of organizational reward systems. *Journal of Leadership & Organizational Studies*, *9*(1), 64–76. doi:10.1177/107179190200900105

Bontis, N., Bart, C., Sáenz, J., Aramburu, N., & Rivera, O. (2009). Knowledge sharing and innovation performance: A comparison between high-tech and low-tech companies. *Journal of Intellectual Capital*, *10*(1), 22–36. doi:10.1108/14691930910922879

Cabrera, E. F., & Cabrera, A. (2005). Fostering Knowledge Sharing Through People Management Practices. *International Journal of Human Resource Management*, *16*(5), 270–735. doi:10.1080/09585190500083020

Carlson, J. R., & Davis, G. B. (1998). An Investigation of Media Selection among Directors and Managers: From "Self" to "Other" Orientation. *Management Information Systems Quarterly*, *22*(3), 335–362. doi:10.2307/249669

Chen, J., Koch, P., Chung, M., & Lee, C. (2007). *Exploring contributory factors in student-to- student knowledge sharing: A Singaporean perspective.* Paper presented at the annual meeting of the NCA 93rd Annual Convention, Chicago, IL.

Connelly, C. E., & Kevin Kelloway, E. (2003). Predictors of employees' perceptions of knowledge sharing cultures. *Leadership and Organization Development Journal*, *24*(5), 294–301. doi:10.1108/01437730310485815

Crossan, M. M., Lane, H. W., & White, R. E. (1999). An organizational learning framework: From intuition to institution. *Academy of Management Review*, *24*(3), 522–537. doi:10.5465/amr.1999.2202135

Cummings, J. N. (2004). Work groups, structural diversity, and knowledge sharing in a global organization. *Management Science*, *50*(3), 352–364. doi:10.1287/mnsc.1030.0134

David, W., & Fahey, L. (2000). Diagnosing cultural barriers to knowledge management. *The Academy of Management Executive, 14*(4), 113–127. doi:10.5465/ame.2000.3979820

Dewett, T., & Gruys, M. L. (2007). Advancing the case for creativity through graduate business education. *Thinking Skills and Creativity, 2*(2), 85–95. doi:10.1016/j.tsc.2007.04.001

DiPietro, W. R., & Anoruo, E. (2006). Creativity, innovation, and export performance. *Journal of Policy Modeling, 28*(2), 133–139. doi:10.1016/j.jpolmod.2005.10.001

Drucker, P. F. (1999). Knowledge-worker productivity: The biggest challenge. *The Knowledge Management Yearbook, 2000–2001*, 266–299.

Fierro, J. C., Florin, J., Perez, L., & Whitelock, J. (2011). Inter-firm Market Orientation as Research. *Business and Management, 4*(1).

Flynn, B. B., Sakakibara, S., Schroeder, R. G., Bates, K. A., & Flynn, E. J. (1990). Empirical research methods in operations management. *Journal of Operations Management, 9*(2), 250–284. doi:10.1016/0272-6963(90)90098-X

Ford, D. P. (2003). Trust and knowledge management: The seeds of success. In Handbook on Knowledge Management (pp. 553-576). Springer-Verlag.

Gardner, H. E., & Laskin, E. (2011). *Leading Minds: An Anatomy of Leadership*. Basic Books.

Gilad, B. (1984). Entrepreneurship: The issue of creativity in the market place. *The Journal of Creative Behavior, 18*(3), 151–161. doi:10.1002/j.2162-6057.1984.tb00379.x

Govindarajan, V., & Gupta, A. K. (2000). Knowledge management's social dimension: Lessons from Nucor Steel. *Sloan Management Review, 42*, 71–80.

Grant, R. M. (1996a). Prospering in dynamically-competitive environments: Organizational capability as knowledge integration. *Organization Science, 7*(4), 375–387. doi:10.1287/orsc.7.4.375

Grant, R. M. (1996b). Toward a knowledge based theory of the firm. *Strategic Management Journal, 17*(S2), 109–122. doi:10.1002mj.4250171110

Grayson, C., & O'Dell, C. (1998). If only we knew what we know: Identification and transfer of internal best practices. *Internal Best Practices, 40*, 154–174.

Gummer, B. (1998). Social relations in an organizational context: Social capital, real work, and structural holes. *Journal Administration in Social Work*, *22*(3), 87–105. doi:10.1300/J147v22n03_06

Hair, J. & Black, W. (2009). *Multivariate data analysis*. Prentice Hall.

Han, B. M., & Anantatmula, V. S. (2007). Knowledge sharing in large IT organizations: A case study. *VINE Journal of Information and Knowledge Management Systems*, *37*(4), 421–439.

Hansen, M. T. (2002). Knowledge networks: Explaining effective knowledge sharing in multiunit companies. *Organization Science*, *13*(3), 232–248. doi:10.1287/orsc.13.3.232.2771

Henri, I. (2016). Review of empirical research on knowledge management practices and firm performance. *Journal of Knowledge Management*, *20*(2), 230–257. doi:10.1108/JKM-09-2015-0336

Hislop, G. W. (2009). The inevitability of teaching online. *Computer*, *42*(12), 94–96. doi:10.1109/MC.2009.411

Hooff, B. V. D., & Huysman, M. (2009). Managing Knowledge Sharing: Emergent and Engineering Approaches. *Information & Management*, *46*(1), 1–8. doi:10.1016/j.im.2008.09.002

Huang, Q., Davison, R., & Gu, J. (2011). The Impact of Trust, Guanxi Orientation and Face on the Intention of Chinese Employees and Managers to Engage in Peer-to-peer Tacit and Explicit Knowledge Sharing. *Information Systems Journal*, *21*(6), 557–577. doi:10.1111/j.1365-2575.2010.00361.x

Huysman, M., & Wulf, V. (2006). IT to Support Knowledge Sharing in Communities: Towards a Social Capital Analysis. *Journal of Information Technology*, *21*(1), 40–51. doi:10.1057/palgrave.jit.2000053

Inkinen, H. (2016). Review of empirical research on knowledge management practices and firm performance. *Journal of Knowledge Management*, *20*(2), 230–257. doi:10.1108/JKM-09-2015-0336

Ipe, M. (2003). Knowledge sharing in organizations: A conceptual framework. *Human Resource Development Review*, *2*(4), 337–359. doi:10.1177/1534484303257985

Kankanhalli, A., Tan, B., & Wei, K.-K. (2005). Contributing Knowledge to Electronic Knowledge Repositories: An Empirical Investigation. *Management Information Systems Quarterly*, *29*(1), 113–143. doi:10.2307/25148670

Kim, S., & Lee, H. (2006). The impact of organizational context and information technology on employee knowledge sharing capabilities. *Public Administration Review*, *66*(3), 370–385. doi:10.1111/j.1540-6210.2006.00595.x

Kogut, B., & Zander, U. (1992). Knowledge of the firm, combinative capabilities, and the replication of technology. *Organization Science*, *3*(3), 383–397. doi:10.1287/orsc.3.3.383

Lee, M. R., & Lan, Y. C. (2011). Toward a Unified Knowledge Management Model for SMEs. *Expert Systems with Applications*, *38*(1), 729–735. doi:10.1016/j.eswa.2010.07.025

Leithwood, K., Jantzi, D., & Steinbach, R. (1999). *Changing Leadership for Changing Times*. Open University Press.

Leonard-Barton, D. (1998). *Wellsprings of knowledge: building and sustaining the sources of innovation*. Harvard Business Press.

Lin, H. F. (2007). Knowledge sharing and firm innovation capability: An empirical study. *International Journal of Manpower*, *28*(3/4), 315–332. doi:10.1108/01437720710755272

Lin, H.-F. (2007b). Effects of extrinsic and intrinsic motivation on employee knowledge sharing intentions. *Journal of Information Science*, *33*(2), 135–149. doi:10.1177/0165551506068174

Ling, C. W., Sandhu, K. K., & Jain, K. K. (2008). Knowledge sharing in an American multinational company based in Malaysia. *Journal of Workplace Learning*, *21*(2), 125–142. doi:10.1108/13665620910934825

Liu, S., & Deng, Z. (2015). Understanding Knowledge Management Capability in Business Process Outsourcing: A Cluster Analysis. *Management Decision*, *53*(1), 124–138. doi:10.1108/MD-04-2014-0197

Loureiro, M. G., Alonso, M. V., & Schiuma, G. (2015). Knowledge and Sustained Competitive Advantage: How Do Services Firms Compete? *Investigaciones Europeas de Dirección y Economía de la Empresa*, *21*(2), 55–57. doi:10.1016/j.iedee.2015.03.001

Lu, L., Leung, K., & Koch, P. T. (2006). Managerial knowledge sharing: The role of individual, interpersonal, and organizational factors. *Management and Organization Review*, *2*(1), 15–41. doi:10.1111/j.1740-8784.2006.00029.x

Mansingh, G., Brysonand, K., & Reichgelt, H. (2009). Issues in Knowledge Access, Retrieval and Sharing—Case Studies in a Caribbean Health Sector. *Expert Systems with Applications, 36*(2), 2853–2863. doi:10.1016/j.eswa.2008.01.031

Marqués, D. P., & Simón, F. J. G. (2006). The Effect of Knowledge Management Practices on Firm Performance. *Journal of Knowledge Management, 10*(3), 143–156. doi:10.1108/13673270610670911

McDermott, R., & O'dell, C. (2001). Overcoming cultural barriers to sharing knowledge. *Journal of Knowledge Management, 5*(1), 76–85. doi:10.1108/13673270110384428

McElroy, M. W., Jorna, R. J., & Engelen, J. V. (2006). Rethinking social capital theory: A knowledge management perspective"". *Journal of Knowledge Management, 10*(5), 124–136. doi:10.1108/13673270610691233

Mesmer-Magnus, J. R., & DeChurch, L. A. (2009). Information sharing and team performance: A meta-analysis. *The Journal of Applied Psychology, 94*(2), 535–546. doi:10.1037/a0013773 PMID:19271807

Mumford, M. D. (2000). Managing creative people: Strategies and tactics for innovation. *Human Resource Management Review, 10*(3), 313–351. doi:10.1016/S1053-4822(99)00043-1

Nahapiet, J., & Ghoshal, S. (1998). Social capital, intellectual capital, and the organizational advantage. *Academy of Management Review, 23*(2), 242–266. doi:10.5465/amr.1998.533225

Nonaka, I., & Takeuchi, H. (1995). *The knowledge-creating company, How Japanese companies create the dynamics of innovation.* Oxford University Press.

Rahimli, A. (2012). Knowledge Management and Competitive Advantage. *Information and Knowledge Management, 2*(7), 37–43.

Riege, A. (2005). Three-dozen knowledge-sharing barriers managers must consider. *Journal of Knowledge Management, 9*(3), 18–35. doi:10.1108/13673270510602746

Rolland, N., & Chauvel, D. (2000). Knowledge transfer in strategic alliances. In Knowledge Horizons: The Present and the Promise of Knowledge Management (pp. 225-236). Buterworth Heinemann. doi:10.1016/B978-0-7506-7247-4.50014-8

Rutten, W., Blaas-Franken, J., & Martin, H. (2016). The impact of (low) trust on knowledge sharing. *Journal of Knowledge Management, 20*(2), 199–214. doi:10.1108/JKM-10-2015-0391

Shalley, C. E., Zhou, J., & Oldham, G. R. (2004). The effects of personal and contextual characteristics on creativity: Where should we go from here? *Journal of Management, 30*(6), 933–958. doi:10.1016/j.jm.2004.06.007

Smith, E. A. (2001). The Role of Tacit and Explicit Knowledge in the Workplace. *Journal of Knowledge Management, 5*(4), 311–321. doi:10.1108/13673270110411733

Suresh, A. (2012). An Empirical Evaluation of Critical Success Factors of Knowledge Management for Organizational Sustainability. *Astitva International Journal of Commerce Management and Social Sciences, 1*(1), 1–12.

Svetlik, I., Stavrou-Costea, E., & Lin, H.-F. (2007). Knowledge sharing and firm innovation capability: An empirical study. *International Journal of Manpower, 28*(3/4), 315–332. doi:10.1108/01437720710755272

Tseng, S. (2008). The Effects of Information Technology on Knowledge Management Systems. *Expert Systems with Applications, 35*(1 & 2), 150–160. doi:10.1016/j.eswa.2007.06.011

van den Hooff, B., Elving, W., Meeuwsen, M., & Dumoulin, C. (2003). Knowledge sharing in knowledge communities. In M. Huysman, E. Wenger, & V. Wulf (Eds.), *Communities and Technologies* (pp. 119–141). Kluwer. doi:10.1007/978-94-017-0115-0_7

Wang, S., & Noe, R. (2010). Knowledge Sharing: A Review and Directions for Future Research. *Human Resource Management Review, 20*(2), 115–131. doi:10.1016/j.hrmr.2009.10.001

Wei, C., The, P., & Asmawi, A. (2012, March). Knowledge Sharing Practices In Malaysian MSC Status Companies. *Journal of Knowledge Management Practice*, (13), 1.

Weiner, R. P. (2000). *Creativity & Beyond: Cultures, Values, and Change.* The State University of New York Press.

Whiting, B. G. (1988). Creativity and entrepreneurship: How do they relate? *The Journal of Creative Behavior, 22*(3), 178–183. doi:10.1002/j.2162-6057.1988.tb00495.x

Williamson, B. (2001). Creativity, the corporate curriculum and the future: A case study. *Futures, 33*(6), 541–555. doi:10.1016/S0016-3287(00)00097-5

Witherspoon, C. L., Bergner, J., Cockrell, C., & Stone, D. N. (2013). Antecedents of organizational knowledge sharing: A meta-analysis and critique. *Journal of Knowledge Management, 17*(2), 250–277. doi:10.1108/13673271311315204

Yi, J. (2009). A measure of knowledge sharing behavior: Scale development and validation. *Knowledge Management Research and Practice*, 7(1), 65–81. doi:10.1057/kmrp.2008.36

Yu, T., Lu, T., & Liu, T. (2010). Exploring Factors that Influence Knowledge Sharing Behavior Via Web Logs. *Computers in Human Behavior*, 26(1), 32–41. doi:10.1016/j.chb.2009.08.002

Zhou, J., & George, J. M. (2001). When job dissatisfaction leads to creativity: Encouraging the expression of voice. *Academy of Management Journal*, 44(4), 682–696.

Chapter 8
Role of Employee Empowerment and Organizational Trust:
Empirical Evidence From the Indian IT Industry

Vandana Singh
Gurukul Kangri University, Haridwar, India

ABSTRACT

This chapter presents the relationship and impact of employee empowerment on organizational trust. The purpose of this study is to analyze the relationship between employee empowerment and organizational trust in the IT industry. This study examines the difference between the empowerment level and organizational level of male and female employees. The questionnaire employed in this study consisted of empowerment by A. K. Mishra and G. M. Spreitzer and organizational trust variables by P. Mishra. The sample for the study consisted of 475 IT professionals from five IT organizations. Simple random sampling was used as a sampling technique, and this study was an ex-post in nature. Data were analyzed using t-test, correlation, and multiple regression. The result revealed that employee empowerment had a positive and significant impact on organizational trust. There is no significant difference in the empowerment of IT industry employees. It means that both male and female employees were equally empowered in their jobs. Male employees are more trust in their jobs as compared to female employees.

DOI: 10.4018/978-1-7998-4180-7.ch008

INTRODUCTION

Employees are the most valuable asset of the organization. Empowerment means encouraging the people to make decisions with least intervention from higher management (Handy, 1993). Employee empowerment starts with the concept of strategic fit between people, tasks, technology and organization structure. Empowered employees depict more trust in their managers. Empowerment practices are often implemented with the hope of overcoming worker dissatisfaction and reducing the costs of absenteeism, turnover and poor quality working condition. Employee empowerment is used as an important tool for understanding of employees about their job role in the organization. Employee empowerment may be subjective by the discernment that the organization thinks about its employees' well-being and that their work is esteemed. Empowering employees may increase job satisfaction. Empowerment influence job satisfaction among employees.

The evolution of trust within an organization will be affected by the history of outcomes that an employee or manager has experienced when trusting the other based on their ability, benevolence, and integrity (Mayer, Davis, & Schoorman, 1995). Organizational trust is a circular motion of action and reaction. Organizational trust is generating increased interest in the organizational studies. Trust has been the centre of the interest of organizational studies since the mid-1980s with the most research emphasizing generalized trust and confirming its importance for organizational development and success (Cummings and Bromiley, 1996). Trust has been widely recognized as a key enabler of organizational success (Davis, Schoorman Mayer and Tan, 2000). Rousseau, Sitkin, Burt and Camerer (1998) proposed the following cross discipline conceptual definition of trust: *"a psychological state comprising the intention to accept vulnerability based upon positive expectations of the intentions or behavior of another"*. From organizational trust (OT) perspective it can be defined as the workers belief that organizations are straightforward in their dealings with workers and follow through their commitments and that the organization's actions will prove beneficial for the workers (Tan and Tan, 2000; and Laschinger, Finegan and Shamian, 2001).

Empowerment

Involving employees in decision making processes, giving them autonomy to complete their tasks and ensuring them that their work has great importance for organization is called employee empowerment (Robbins, P. S & Judge, A., 2000). Employee empowerment means sharing the information about vision, clarity of goals and defining the borders of decision making process (Nanda, N & Nanda,

I., 2009). The construct of empowerment comprise of four dimensions meaning, competence, self-determination and impact.

- **Meaning**: Meaning is an individual's belief that his or her work is important to him or her and his or her fondness for what he or she is doing. Spreitzer (1995) defined meaning cognition as a sense of purpose or personal connection to one's work goal: it is the value of a work goal and an individual's beliefs, judged in relation to that individual's ideals (Thomas and Velthouse, 1990).
- **Competence**: Competence is an employee's ability to perform work activities with skill, and it refers to the degree to which a person can perform task activities skillfully when he or she attempts to do so. Employees must feel that they are competent to engage in the behaviors required by the environment (Kara, 2012).
- **Self-determination:** Self-determination is an individual's control over the manner in which work is accomplished and is related to the choice cognition, as described by Kitayama and Cohen (2010). Such individuals feel that they have the independence to perform their responsibilities; they can make decisions about their work and have adequate authority over the manner, time and speed of their task performance (Hossein et al., 2012).
- **Impact**: Impact is the degree to which an individual can influence strategic, administrative and operating outcomes at work. This assessment refers to the degree to which a behavior is viewed as making a difference in terms of accomplishing the purpose of the task that is, producing the intended effects in one's task environment (Thomas and Velthouse, 1990).

ORGANIZATIONAL TRUST

Lewis and Weigert (1985) understood trust as the undertaking of a risky course of action with the confident expectation that all persons involved in the action will act competently and dutifully. According to Mishra (1994) Trust is one party's willingness to be vulnerable to another party based on the belief that the latter party is competent, open, concerned, and reliable.

Dimensions of Organizational Trust

Organizations cannot function without the people who, although they are individuals depend on others. This interdependence requires collaboration, which is only successful if it is based on trust. Trust is at the heart of an organization's ability to succeed and the ability to have confidence in relationships of all types is critical.

Research suggests that the core elements of trust include openness, concern, working condition, reliability, communication and relationship. Although not all elements are equally important in all situations and their importance varies dependent upon the scenario, it is accepted that they are all essential to healthy and productive relationships. They are openness/honesty, concern for employees, reliability and identification.

Many believe there is nothing we can do to actively build trust, but we disagree. Everything we do is about trust-building, and communicators are at the forefront of trust-building strategies in organizations. Strategic communication makes all the difference. Trust is the main thing.

REVIEW OF LITERATURE

Moye & Henkin (2006) explored association between employee empowerment and interpersonal trust. Regression analysis revealed that, employees who felt empowered in their work environment tended to be at higher levels of interpersonal-level trust with their managers. Ergeneli, Ari, & Metin (2007) examined the relationships between overall psychological empowerment, as well as its empowerment four dimensions (Spreitzer): meaning, competence, self-determination, and impact. Correlation results showed a significant relationship cognition-based trust in immediate managers between the overall psychological empowerment. Although cognition-based trust related with the meaning and competence dimension, affect-based trust related to impact only. With demographic as the control variables, only position had an impact on the psychological empowerment. Findikli, Gulden, & Semercioz (2010) examined the relationship between psychological empowerment and trust in the organization and the supervisor. There was no significant difference in terms of position, experience and age, whereas, given trust in supervisor, there was a significant difference in terms of gender and education. The findings revealed that there was a strong and positive relationship between trust in supervisor and the subordinate perceptions of the psychological empowerment in terms of meaning and competence factors. Trust factor was an intervening variable thought to be an effect on the perceptions of psychological empowerment. Knoll & Gill (2011) investigated the generalizability of the Integrative Model of the organizational trust. The correlation and multiple regression results revealed that the integrative model of organizational trust was applied to trust in supervisor, subordinate and peer. Results suggested that the relative importance of ability, benevolence, and the integrity in predicting trust differed according to the trustor-trustee dyad. Judeh (2012) explores the present status of inter-organizational trust and employee empowerment. Correlation and regression results indicate that trust was at a medium degree & employee empowerment was at a high level. There was a significant correlation between organization trust and employee empowerment. According to the result of the regression analysis, interpersonal and organizational trust had a close relationship with all dimensions of empowerment. Alizadeh & Panahi (2013) Findings of correlation revealed that there was a strong & positive relationship between the organizational trust and dimensions of organizational culture. Mangundjaya (2014) results showed that both psychological empowerment and organizational trust had a positive and significant correlation and contribution to commitment to change and both the organizational trust and psychological empowerment had higher the impact to affective commitment to change compared to other normative and continuance dimensions of commitment to change. Erturky & Vurgun (2015) results indicated that organizational trust and empowerment and empowerment dimensions positively correlated with them. Alajmi (2017) results indicated that a significant positive correlation exits between empowerment of the employees of these companies and organizational trust. And with all empowerment dimensions significantly correlates with organizational trust. Mollamohammadrafie (2019) result indicated that empowerment and empowerment dimensions had a significant and positive relationship and impact on organizational trust.

OBJECTIVES OF THE STUDY

1. To study the effect of personal variables on empowerment and organizational trust of professionals in IT industry.
2. To analyze the relationship between empowerment and organizational trust of professionals in IT industry.
3. To examine the influence of the empowerment (meaning, competence, self-determination and impact) on organizational trust of professionals in IT industry.

HYPOTHESES

$H_0 1$: There is no significant difference in organizational trust of IT industry professionals on the basis of Gender.

$H_a 1$: There is significant difference in organizational trust of IT industry professionals on the basis of Gender.

$H_0 2$: There is no significant difference in employee empowerment of IT industry professionals on the basis of Gender.

$H_a 2$: There is significant difference of employee empowerment of IT industry professionals on the basis of Gender.

$H_0 3$: There is no significant relationship between employee empowerment and organizational trust.

Ha3: There is significant relationship between employee empowerment and organizational trust.

$H_0$4: There is no significant impact of empowerment (meaning, self-determination, competence and impact) on organizational trust.

Ha4: There is significant impact of empowerment (meaning, self-determination, competence and impact) on organizational trust.

METHODOLOGY

Sample and Sampling Technique

Sample for the present study consisted of the 500 IT Professionals of Delhi NCR Region, India. The simple random sampling technique was used to select the sample. A total of 500 questionnaires were distributed. The respondents were allowed to take their own time in filling the questionnaire. Out of 500 questionnaires, 490 questionnaires were received back, giving a positive response rate of 98% out of which 475 were found usable in the study.

Data Analysis

The Statistical Package of Social Science (SPSS) was used for analysis Independent Sample t-test, Correlation and Regression were used in this paper.

ANALYSIS AND INTERPRETATION

Table 1 shows the results of t-test on organizational trust on the basis of gender of the respondents in IT industry. A significant difference (at 0.05 level of significance) was found between male and female employees. Since $p = .017 < 0.05$, therefore, the null hypothesis is rejected. Therefore, it can be inferred that gender made significant

Table 1. T-test Statistics for Organizational Trust on the basis of Gender

Gender						
		N	Mean	Std. Deviation	t-Test	Sig.
Organizational Trust	**Male**	365	71.33	13.762	1.236	.017
	Female	110	69.71	11.485		

*significant @ 0.05 level

difference in the organizational trust of IT industry employees. It means that male employees (mean = 71.33) had more trust as compared to female (mean = 69.71) employees in their jobs.

Table 2. T-test Statistics for Empowerment on the basis of Gender

Gender						
		N	Mean	Std. Deviation	t-Test	Sig.
Empowerment	Male	365	54.91	9.095	2.103	.520
	Female	110	52.85	8.548		

*significant @ 0.05 level

Table 2 shows the results of t-test on empowerment on the basis of gender of the respondents in IT industry. A significant difference (at 0.05 level of significance) was not seen between male and female employees. Since p = .520 > 0.05, therefore, the null hypothesis is accepted. Therefore, it can be inferred that gender (male and female) made no significant difference in the empowerment of IT industry employees. It means that both male (mean = 54.91) and female (mean = 52.85) employees were equally empowered in their jobs.

Table 3. Coefficients of Correlation between Empowerment Dimensions and Organizational Trust

Dependent Variable		Meaning	Competence	Self-determination	Impact	Empowerment
Organizational trust	Pearson Correlation	.493	.424	.519	.720	.769
	Sig. (2-ailed)	0.001	0.001	0.001	0.001	0.001
	N	475	475	475	475	475

** Correlation is significant at the 0.01 level (2-tailed).

Table.3 shows the results of correlation analysis between the dimensions of empowerment and organizational trust in IT industry. The results indicate that there is a positive and significant correlation between dimensions of empowerment and organizational trust. The relationship of organizational trust with empowerment dimension impact (r = .720, p < 0.01) is high followed by self-determination (r =

Table 4a. Multiple Regression Analysis

Model Summary				
Model	**R**	**R Square**	**Adjusted R Square**	**Std. Error of the Estimate**
1	.773 [a]	.598	.597	6.879

a. Predictors: (Constant), Empowerment

Table 4b. Multiple Regression Analysis

ANOVA[a]						
	Model	**Sum of Square**	**Df**	**Mean Square**	**F**	**Sig.**
	Regression	33259.006	1	33259.006	702.901	.000 [b]
1	**Residual**	22380.838	473	47.317		
	Total	55639.844	474			

a. Dependent Variable: organizational Trust
b. Predictors: (Constant), Empowerment

.519, $p < 0.01$), meaning ($r = .493$, $p < 0.01$) and competence ($r = .424$, $p < 0.01$). The results indicate that there is a positive and significant correlation between empowerment and organizational trust ($r = 0.769$, $p < 0.01$).

It is concluded that the high organizational trust leads to high empowerment amongst employees in IT industry.

Table 4 presents a multiple regression analysis where organizational trust was regressed on empowerment of IT industry. The result shows that empowerment made a significant contribution to the prediction of organizational trust of IT industry. It can be inferred that the empowerment of employees contributes positively towards the IT organizational trust. Thus, empowerment have significant impact on organizational trust of the IT sector.

Table 4c. Multiple Regression Analysis

Coefficients[a]						
Model		**Unstandardized Coefficients**		**Standardized coefficients**	**T**	**Sig.**
		B	**Std. Error**	**Beta**		
1	(Constant)	9.453	2.677		3.531	**.000**
	Empowerment	1.155	.044	.773	26.512	**.000**

a. Dependent Variable: Organizational Trust

DISCUSSION

The literature suggests the empowerment has an expressive role in many service organizations, including IT sector. The present study examined the effect of employee empowerment on organizational trust in IT sector. Four dimensions were used to assess employee empowerment: meaning, competence, self-determination and impact. On the basis of analysis, it is concluded that employee empowerment has positive and significant impact on organizational trust and the findings were consistent with the findings of the studies conducted by Mangundjaya (2014) study showed that both psychological empowerment and organizational trust had a positive significant correlation and significant and positive impact of empowerment on organizational trust in organization. Further it has been found that there is significant difference among gender base on organizational trust level. The finding of significant difference among gender on organizational trust was also consistent with the work of Findikli, et al. (2010) revealed that gender had significant difference in organizational trust. And there is no significant difference among gender on empowerment level. It states gender made no significant difference in employee empowerment. This study confirms that employee empowerment leads towards higher level of organizational trust and findings indicate that the dimensions of empowerment i.e. meaning, impact and self-determination and competence have significant influence on organizational trust. Employees who perceived themselves as empowered felt that they had an influence on their daily work, were independent, had autonomy with regard to the manner in which they performed their jobs and were proud of their jobs. The study has supported this version of empowerment and organizational trust.

LIMITATIONS AND FUTURE RESEARCH

This study has taken the IT professionals from Delhi NCR-Region, so, views and thoughts of these professionals may not reflect the views and thoughts of all the professionals of entire NCR-Region and India which limits the scope of this study. Additionally, this study concentrated on IT employees so future research could extend the investigation to different sectors to obtain a wider generalization of the study. Future studies can also be tailored to investigate the employees view point of empowerment and organizational trust of IT professionals. For the purpose of connection, it would be interesting to replicate this study in a longitudinal design, so that it could be determined if empowerment and organizational trust are conditions and relationships that are likely to be sustained.

CONCLUSION

This study contributes to overcome the problems of employee empowerment in IT industry and give a fair idea that employee trust can be achieved through empowerment. The present study will help to improve the process of empowering employees in IT sector of Delhi NCR region. Almost all firms had recognized the importance of increased employee performance, satisfaction and trust for organizational sustainability and development. This is possible only by empowered workforce. Since we have found employee empowerment as an important factor that enhances employee trust, it is recommended that further studies should be carried at the higher level with larger sample size, more demographics factors must be added in the study and should be expanded to the IT industries throughout the country.

REFERENCES

Alajmi, S. A. (2016). Organizational trust: A gateway to psychological empowerment. *Journal of Management Research*, *9*(1), 52–69.

Alizadeh, S., & Panahi, B. (2013). Organizational culture constructs in the development of organizational trust. *International Journal of Management Research and Review*, *3*(8), 3238–3243.

Cummings, L. L., & Bromiley, P. (1996). The organizational trust inventory (OTI): Development and validation. In R. Kramer & T. Tyler (Eds.), *Trust in organizations* (pp. 302–330). Sage.

Davis, J. H., Schoorman, F. D., Mayer, R. C., & Tan, H. H. (2000). The trusted general manager and business unit performance: Empirical evidence of a competitive advantage. *Strategic Management Journal*, *21*, 563–576.

Ergeneli, A., Ari, G. S., & Metin, S. (2007). Psychological empowerment and its relationship to trust in immediate managers. *Journal of Business Research*, *60*(1), 41–49.

Erturk, A., & Vurgun, L. (2015). Retention of IT professionals: Examining the influence of empowerment, social exchange, and trust. *Journal of Business Research*, *68*(1), 34–46.

Findikli, M. A., Gulden, A., & Semercioz, F. (2010). Subordinate trust in supervisor and organization: Effects on subordinate perceptions of psychological empowerment. *International Journal of Business and Management Studies*, *2*(1), 55–67.

Handy, M. (1993). Feeling the victims. *Total Quality Management*, 11.

Hossein, R. D., Saleh, P. A., Iman, A. M., & Jaafar, A. (2012). An analysis of the empowerment level of employees and it's relation to organizational factors. *International Journal of Business and Social Science*, *3*(15), 255–263.

Judeh, M. (2012). An analysis of the relationship between trust and employee empowerment: A field study. *International Business Management*, *6*, 264–269.

Kara, D. (2012). Differences in Psychological Empowerment Perception of Female Employees Working in Hospitality Industry. *Middle East Journal of Scientific Research*, *12*(4), 436–443.

Knoll, D. L., & Gill, H. (2011). Antecedents of trust in supervisors, subordinates, and peers. *Journal of Managerial Psychology*, *26*(4), 313–330.

Laschinger, H. K. S., Finegan, J., Shamian, J., & Wilk, P. (2001). Impact of structural and psychological empowerment on job strain in nursing work settings: Expanding Kanter's model. *The Journal of Nursing Administration*, *31*(5), 260–272.

Lewis, J. D., & Weigert, A. J. (1985). Trust as a social organization. *Social Forces*, *63*(4), 967–985.

Mangundjaya, W. L. H. (2014). People or Trust in Building Commitment to Change? *Proceedings of the Australian Academy of Business and Social Sciences Conference*.

Mayer, R. C., Davis, J. H., & Schoorman, F. D. (1995). An integrative model of organizational trust. *Academy of Management Review*, *20*(3), 709–734.

Mollamohammadrafie, H. (2019). The effect of psychological empowerment and organizational trust on affective commitment evidence from Padjadjaran University, Bandung Indonesia. *International Journal of Scientific & Technology Research*, *8*(1), 63–69.

Moye, M. J., & Henkin, A. B. (2006). Exploring associations between employee empowerment and interpersonal trust in managers. *Journal of Management Development*, *25*(2), 101–117.

Nanda, N., & Nanda, I. (2009). Employee Empowerment and its Organizational Effectiveness: A study on Rourkela Steel Plant, Odisha. *India International Journal of Research in IT. Management and Engineering*, *1*(3), 1–8.

Robbins, S. P., & Judge, T. A. (2000). *Organizational behaviour*. Prentice Hall.

Rousseau, D. M., Sitkin, S. B., Burt, R. S., & Camerer, C. (1998). Not so different after all: A cross-discipline view of trust. *Academy of Management Review, 23,* 393–404.

Spreitzer, M. (1995). Psychological Empowerment in the Workplace: Dimensions, Measurement, and Validation. *Academy of Management Journal, 38*(5), 1442–1465.

Tan, H. H., & Tan, C. S. F. (2000). Toward the differentiation of trust in supervisor and trust in organization. *Genetic, Social, and General Psychology Monographs, 126,* 241–260.

Thomas, K. W., & Velthouse, B. A. (1990). Cognitive elements of empowerment, an interpretive model of intrinsic task motivation. *Academy of Management Review, 15,* 666–681.

Chapter 9
Technology Applications in Managing Talents

Pushpa Kataria
Doon Business School, Dehradun, India

ABSTRACT

Human resource (HR) management is all about people; there is no doubt about it. However, in the contemporary era inbuilt by the technological revolution and transformative development, the tenets of HR are finding a new footing. Technology applications have changed facets of corporate houses from restructuring organizations to resources to manpower. It has been seen that HR technologies are playing a major role in managing talents in organizations. Artificial intelligence, robotic process automation, and machine learning add an edge to talent acquisition and management. Machine learning tools are primarily being used in acquiring talents and enabling the hiring process effectively and efficiently. Today, as every company is moving to a new era of digitalization and data management, managing and mapping talents pose a challenge to C-suite and board-level management. The present study highlights the role of technology in managing talents in a series of HR processes.

INTRODUCTION

Human resource (HR) management is all about managing human assets in the organization which meet end to end process from procuring to retaining of human resources. However in the age of digitalization, human resources management has undergone a paradigm change to meet technological innovations in the organizations. Innovation and technology have changed facets of corporate houses from restructuring organizations to resources to manpower. Technological transformation is playing a key

DOI: 10.4018/978-1-7998-4180-7.ch009

role in augmenting employee lifecycle management and learning and development at organizational level. (Davenport et al., 2010; Snell, 2008). Today, the journey of human resources management, from talent acquisition, management, training candidates to retaining, is influenced by the tools of technology. This makes pertinent HR function to keep updated with every technological aspects and its applications to job roles. Technology applications have changed facets of corporate houses from restructuring organizations to resources to manpower. It has been seen that HR technologies are playing a major role in managing talents in the organizations. Artificial Intelligence, Robotic Process Automation and Machine Learning adding an edge to talent acquisition and management. These technologies are facilitating core HR functions of Acquiring, developing, maintaining and retaining of talents in the organizations. Machine learning tools are primarily being used in acquiring talents and enabling hiring process effectively and efficiently.

Today, as every company is moving to a new era of digitalization and data management, managing and mapping talents poses a challenge to C-suite and board level management. Therefore it is vital for a corporate house to recognize emerging technologies which impacts the dimension of talent management (Hashani & Bajrami, 2015). As per the latest report submitted by Deloitte (2020), nearly 42% of the organizations either have fully established or made important advancement in adopting cognitive and artificial intelligence technologies within their talent force. The newest corporate mantra which is redefining and inventing human processes include big data, machine learning, mobile applications, social media, online onboarding, cloud technology, and business models like SaaS business model. The concept of disruptive innovation and automation bringing tremendous transformation in organizational efficiency and effectiveness, in line with the compliance regulations. Technology has absolutely changed the facade of human resources universally, from administrative roles in the organization to the most critical role of a strategic partner in competing growing businesses. According to US survey conducted by Society for *Human Resource* Management (SHRM) in 2018, found that 77% leaders in the field of human resources utilized analytics to analyse and communicate with employees, while 79% of them used automated HR solutions to carry out more planned roles and decisions in the organization. As human resources is more oriented towards ensuring employees success and contribution, it can be enabled to strategic dimension and actions of an organization. Technological tools in the field of human capital management supports multiple operations, including sub dimensions of recruitment, payroll management, manpower inventory tracking etc. Novel organizations have also come up with assessment centres to identify and align human skills with technological moves and their adaptability for being more agile.

TALENT PLANNING AND MANAGEMENT

Talent planning is defined to be too crucial phase for an organization in making the right strategy alignment. Smart companies relies on SAP Success Factors, Workday, Oracle Hyperion Workforce Planning etc., though there are many existing business models like Human Capital Institutes, Strategic Workforce planning but still provides a gap analysis.

Talent management is the organized practice of "seeking the vacant position, acquiring the suitable person, budding the skills and proficiency of the person to match the job profile and retaining him/her to achieve business objectives (Blass. E., 2007) . The talent management process aligned with organizational strategy achieves the objectives of performance outcomes in an organization. In other way it can be concluded that talent management process is the function of organizational performance outcomes. The talent management process model in Figure 1, identifies and utilizes technology at each step of manpower planning, employee sourcing, on- or off-boarding, performance evaluation, training & support, career planning, reward management and skills gap analysis to workforce diversity. (Tubey, R., Rotich, K. J., & Kurgat, A.,2015). The managers and the core human resources need to take the HR initiatives to fit technologies in talent planning to talent control in the organisation.

Companies are using technology facilitation into people development for example-Talent Edge is utilized in Infosys and PeopleSoft at Oracle and TCS has talents programs like CLP-Continuous learning programme, ILP-Initial learning programme, FLI-Foreign language initiatives, LDP-Leadership developing programme.

Figure 1. Talent Management: A Process

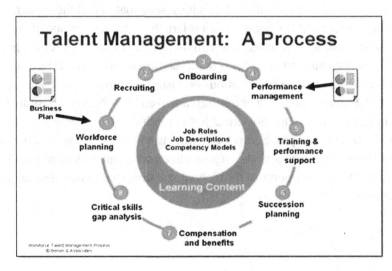

Contemporary industries are looking for a novel technology approaches like Textio, which helps hiring firms to assess the performance of their JD in comparison to the same with millions of JD's. Further a tool called Resume Assistant, brought, powered by LinkedIn helps a candidate to write an approachable resume

TECHNOLOGICAL INFLUENCE ON TALENT ACQUISITION & MANAGEMENT

It is noticed from past surveys that technology and talent acquisition goes hand in hand (Bawany, 2018). With the technological disruptions, hiring process in the organizations have underwent a lot of change. To go back, few years ago, acquiring human resources was based on the employees' credentials where an employee was selected purely based on the academic certificates, experience or other qualities as per job requirement. But now, technology has changed the things completely. Today a candidate is judged more efficiently and quickly with the invention of technology based platforms(Guer et al., 2012). The tech enabled systems help organizations to receive mass applications, sorting and shortlisting candidates as per the job requirement of the company. Choosing a right technology which facilitates HR processes in the organizations aims to achieve organizational objectives more effectively (Bal et al., 2012). For example an IT based company at Bengaluru named Mphasis, running an app called 'Ask Dexter' which is successfully able to manage more than 22,000 workforce in the organizational setup. The company is using internal, cloud-based Chabot, in controlling HR processes. The system is very well built to resolve the employee queries on types of leaves and organizational policies. Its technical support system also provides employee appreciation at cross functional level in the organization. It has been seen that one of the modern source of acquiring resources in the organization is largely influenced by social media. HR Managers/ professionals are using powerful social media tools in recruiting people as it provides an open and enlarged basis to search for candidates and the information available on social media. Organizations are using simple and precise recruitment process via Facebook, LinkedIn and Instagram to contact the right mass at the right time and right place. Companies are able to search capable people more with the advent of internet based process and it is proving to be one of the effective way of attracting talented resources to the organizations who can understand technology very well (CIPD, 2006b). Some of the technological application in talent acquisition and management are as follows:

BIG DATA POWER ORGANISATIONS

Compliances and controlling risks are the two key powerful aspects of every functional role of a department. Morever, technological applications have changed the droning and routines jobs via concept of HR tech and information based HR tools and systems. The most widely used technological tool is Big Data, enabling HR professionals to connect organizational customers, markets and services in the most satisfy ways. Integrated and merged with original technologies, Big Data aids to expand in-depth insights and rolling up HR professionals to formulate decisions.

Big Data is an effective tool that provides HR manager with a proof-based outlook of the present workforce in the organization and enable them to recognize upcoming trends on HR processes. Furthermore, analytics assists recruiters/ consultants to assess prospective competencies of employees and improved management decisions.

MOBILE APPS - NEW NORMAL

Smartphones is another vibrant zone that is also leading the HR dimension. It has been noticed that tech workforces in various operations/ cross functional areas filling applications through mobile and other gadgets, business houses are beckoning on integrated HR systems. Adopting such kind of activity by the organizations implies that organizations have to depend on HR applications supported by mobilization and the friendly interface that employees usually look at. As understood, the tendency of applying through mobile apps continues, the basic HR functionality on technological support shall continue to depend. Mobile applications have become the most common platforms for application submission and interaction.

SOCIAL MEDIA – A POWERFUL INTERACTIVE TOOL

Social media platforms have occupied a special place in development and facilitation of talent management today, essentially when it comes to recruitment. Almost all modern employers/companies are exploring social medias such as Facebook and LinkedIn to pitch competent employees. HR departments are not only using social media in talent acquisition but also in managing HR systems and processes. Hence, social media is understood as a chief foundation for companies to attain their HR objectives. Corporate are using social media channels to focus on right candidature with the help of job postings. Most of the corporates have made social media platforms a vital part of their organizational structure to convey their organizational success stories via medias like videos, photographs, blogs, Tumblr, and also Pinterest pages. Job seekers specially inexperienced or campus fresher's are using social networks to know about the company and other vital information. Social medias are excellent source to explore companies feedback through current employees and customer reviews.

Social media brings abundant paybacks to human resource professionals and it can be best be mingled with information, technology, and trends. It helps to nurture employees' relationships by sharing industry knowledge sources and information. Because of all these benefits, social media have become an ideal stand to keep more engaged employees, build harmonious relationships, and generate relational power communications in the organizational setup. Therefore companies have to integrate applications with LinkedIn or Facebook than to developing applications.

CLOUD TECHNOLOGY AND SAAS

Another important trend based on web or native application is SaaS apps which has already emerged in every sector, with HR as no exception. Cloud-based applications is growing and complementing today's business environment. Data storage and Anthology remained as one was one of the biggest bottleneck till the era of development of the cloud. With the growth of cloud technologies, storage of information, accessibility of documents and other important data, has become easy mode of availability through online mode. Cloud computing has made data collection and organization in a most protected way.

At the same time it is essential to understand cloud-based solutions to fit to the technology used by the company's current system and processes and it should upgrade the genuine value adding to the company's business. Organizations need to understand and compare the potential challenges of cloud based solutions against the profit reached through cloud that helps to overcome business risks. It is pertinent to notice that business procedures to merge with cloud applications. Centralization of data/information or workflows and operations can be better aligned with the organization. Cloud solutions must lead to positive influence on various operations from stages of product expansion, personnel management to business growth and integration. It can be concluded that most of the organizations are shifting to cloud-based applications.

BRING YOUR OWN DEVICE (BYOD)

Most companies adapt to latest innovation and technological practices that helps to make their products work smarter. BYOD has become a novel practice today, and companies have started embracing this practice due to its huge success to the company in terms of low investment and cost. BYOD creates savings on the part of the companies as they do not required to provide workforce with the hardware, infrastructure and other systems for work. It has been identified from various literatures that employees are successfully able to fulfill their personal and professional lives with full contentment when they utilizes self devices for completing work tasks.

Many multinational companies are have accepted this trend and allowing the BYOD practices, even from the security point of view. However, companies are ensuring the application of BYOT program in place complying with their policies for securing confidential data or also in cases when employee device is lost or stolen.

WEARABLE TECHNOLOGY

Some of the example of companies like Google, Apple, and Microsoft have introduced wearable technology in all their devices at workplaces. The wearable devices helps in improving motivational level of employees and build connection with each other. Wearable devices lead to improve time management. Wearable technologies have been discovered to offer great opportunities for HR processes and employee engagement, enhance productivity, benefit incentives, and security issues (Glen, and Clayton, 2006). Wearable devices build communication system and functional efficiencies, businesses are exploring new avenues to create wearable device into a powerful tool. Organizations are planning to switch to wearable devices to gather information.

BLOCKCHAIN- BRINGS NEW SECURITY

Today most of the organizations are protecting their HR records by making use of block chain. Blockchain helps an organization in maintaining and safeguarding HR records without any catch of virus or data leak. It transparently maintains HR data with no issues of data tampering.

INTEGRATED TALENT MANAGEMENT: A NEW PARADIGM IN HR

Hence it can be concluded that with the rapid expansion of technology has given a rise in the growth of integrated talent management system. It is seen that HR departments are no longer traditionally independent and documented but it is going for an integrated talent management systems that bind together the diverse functionalities of HRIS & managing talent (Dwivedi, Kiang, Williams, and Lal, 2008). As a single integrated system, companies are targeting at amalgamation and examination of the core HR data to reach at strategic decisions- obtaining, developing, evaluating and training/upskilling of talents (Biswas 2019).

TALENT ACQUISITION AND INNOVATIONS

A successful talent acquisition is essential to the success of an organizational stability (Dery, Hall, & Wailes, 2006b). Therefore many organizations are now a day's using innovative technologies recruiting talents at various positions. Machine learning, algorithms and AI tools are being utilized at various process of recruiting.

Job Adverts

Job adverts (any other type of text for that matter) are the novel HR innovation. This augmented writing technology help companies in acquiring the right candidate. This tech machine learning and predictive analytics uses machine learning and predictive analytics that is able to target the candidates through the use of augmented writing technology which increases the chances of getting a desirable outcome.

Sourcing of Candidates

AI tools are also extensively used in the sourcing talents are from within and outside the organization (Gakovic, Anika and Keith Yardley, 2007). The systems are created in such a manner that sometimes machine automatically contact an employee and would do the preface short listing by using chatbots to in identifying the right talents. Video Interview Bots are also utilized in automatically schedule, arrangement of interviews and even in assessment of candidates based on standardized criteria. For example, programmatic job advertising ensures, organization's cost per application at reduced time per hire.

Chatbots

Chatbot is a powerful device of software – At times powered by artificial intelligence that can handle a several tasks at a time. For example sourcing, screening and arrangement of candidates get ease with chatbots. Chatbots also have inbuilt Natural Language Processing (NLP) capabilities that is very well 'understand' human language. Chatbots is a meritorious concept as it is found never out of office(as human may go home in the evening and during the weekdays but a chatbot never rest or stop) able to handle responses of candidate questions any time without holding a candidate.

Preselection Technology

Preselection technology are enabling organizations to hire candidates selectively by adding right talents to the boards. Preselection is a Human Resources best practice that augments profitability of the company. Pre-employment assessment software go through dozens of CVs or sometimes even thousands of CVs. For example companies are using AI-powered multi-assessment platform for initiating CVs.

On-Boarding

Automation of onboarding process, is the present happening stories in emerging organizations. SAP Success Factors is an AI-powered "Onboarding buddy," that facilitates companies to set up certain answers to typical specimen questions that employees may come up on joining. AI-powered technologies helps to make talent decisions in effective manner and help companies in simplifying talent activities and processes especially for onboarding talent. Today, Artificial Intelligence (AI) helps for more refined search (Mark Skilton, 2010). AI usage for sourcing, screening and interviews is one of the ways to improve the candidate experience while undergoing talent screening. Hilton, Humana, Thredup are some of the companies that are using AI in hiring employees at workplace. According to Bloomberg and Dolby Laboratories (2016) have explores Gap- Jumpers to diminish prejudices in recruiting candidates.

Online on Boarding

Online onboarding is a helpful process when used at cross-location, it standardizes the material contents and avoids information overloaded to the new joined employees. Some of the advantages derived out of online onboarding helps in easing out:

- Online paperwork/formalities completion before a candidate joins an organization.
- An E-employee handbook can also be shared before to the joining of the employees.
- An E-mentor / Buddy can be allotted to begin a social connect
- Arrangement of E-learning modules for the candidates
- videos or podcasts about the brand and culture of the company can be shared easily

- Social engagements of new hire to via external and internal social channels can be made quickly

The onboarding process must be necessarily be complemented and supplemented by a personal touch that enables the new hires feel esteemed. Therefore identifying the right onboarding areas and employing using both offline and online channels help in keeping employees engaged in the organizations. Most of the companies are using a data and metric-driven methodology to onboarding that can measure the effectiveness and efficiency of onboarding process at various stages. Measurement can be carried in two ways:

- Gauging employee progress: Metrics can be employed for such as half-yearly retention, annual retention, new hire evaluation etc. can be tracked easily and can be associated to onboarding.
- It can be focussed to process execution such as training, performance statistics.

Truthful Predicting and Leveraging Technology

Another important key component of effective agile TA embracing is the capability to estimate future talents demand forecasting and determining pivotal roles that will augment business performance. Though data gathering and methodologies are being used popularly, at the same time it is relevant to leverage them tactically to make effective decisions. Therefore this course of forecasting and forecasting talents prerequisites allows a blend of agile talent acquisition and workforce planning in the organizations. It is an important process that will help many TA heads/sources to initiate vital decisions in addressing best decisions for talent demands for their companies comprising of permanent, contingent, contractual, and cyclic workers.

When it comes to accurately forecasting personnel needs, HR specialist needs to leverage the state-of-the-art in data, analytics, know-how and technological innovations. Venture in AI-technologies can endow the data assembling and enabling the free time for the recruiters. Technology and AI have proved to ease process such as sourcing, aids recruiters more timely efficient and agile. It also help to find the right candidate thus helping recruiters to make decisions more agile and enabling TA to be fasters.

Implementing agile practices, today, has majorly influenced corporations and organizational functionalities through its original origin. Today almost all contemporary organizations are using tech-driven business renovations, with innovation becoming the taking the cradle of HR practices in 21 century. TA models are greatly benefiting the firms who are undergoing a rapid change as per global trends. Using the facility of AI to complement (not to replace) human judgements can make the Talent Acquisition procedure simpler.

Making an informed choice are often the primary step in having happy employees.

As technological capabilities increase, there must also be a system to upgrade citizenry to maneuver beyond the restrictions of their intellect and are available to a deeper dimension of intelligence.

Technology will bring comfort and convenience but won't bring wellbeing. It is time to focus on inner wellbeing. AI is not set to place HR Managers out of labor , rather it is set to rework HR systems across entire organization. Technology is growing a robust game changer, hence creating transformation in business place. In current generation, human and Artificial Intelligence (AI) is creating wonders. It is evident that AI is not set to replace HR out of work, rather it is meant to transform HR processes and culture of operation..

Therefore, it is obvious most organizations are moving towards adaptation of AI in improving business decisions not leaving behind HR activities. KPMG in its International's latest global survey(2019) indicated that 60% of HR executives felt "very confident" about HR's original ability to rework and move upward through key skill set like HR analytics and AI.

Talent acquisition, for example, has been understood as key area for an HR concern (Blass, 2007; CIPD, 2008; Snell, 2008). Most organizations need to ask questions of how do they ensure "what we have hired is indeed talent?" Sometime a street-smart employee may fails to live up to potentials, where AI can help to restructure, processes in line with analytical processing of data.

AI helps in designing questions for round of employee interviews which do not take into account a person's background at the cost of their competency for a job role. The questions are based on job requirement at organizations rather than based on an individual's previous work records. AI's assistance in talent management helps in making successful talent creation and retention.

AI helps in mapping employee talents within minutes, aiding HR managers in implemetation of accurate decisions. In India, AI shall drive significant value in HR, but still long way to go.

KPMG International survey (2019) revealed that more than 50% of HR functions have introduced AI and around only 26% were able to invest in AI in the past.

This indicated that India has marked with a series of dichotomies, KPMG (2019). HR leaders have shown a great strength to strive in the direction of transforming businesses and HR functions as equivalent to global counterparts. It is noted that HR departments in organizations is one of the foremost promising field to place AI to all primary lines in handling the "human" component of their businesses to grievances management. AI, if implemented in right way, could function an excellent ally at every stages-sourcing talents, short listing applicants resumes, screening applicants, on-boarding and performance assessment process, making it a successful tool in HR management.

Gig Economy

According to an article, Peoplematters (2020), Gig-economy is emerging as a contemporary workforce and is impacting the global platform, in a more diversified, agile, handy and interesting, Many MNCs are leveraging on gig workers to ascertain the strategic outcomes of the organizations (Berg et al., 2018). Though the concept of gig workers have been prevailing in the past but due to the improved technology and structures at workplaces, they are developing as an important division of the workforce.

CONCLUSION

There has been many technological innovations in talent acquisition and management augmenting end to end HR processes, thereby making AI powered organizations. Most of the companies using technology tools like Do It Yourself (DIY) mode, where recruiters are focusing on employer brand and rapport building exercises.

AI has been termed as the "game changer" with in-depth potentialities to redesign, reinvent and reform various areas of talent acquisition to talent management and so on. However, many organizations are still reforming to leverage this powerful technology to get-up-and-go talent competences. Companies are forwarding to invest in AI, machine learning, cognitive technologies, robotics, and automation.

As seen with world champions and world of changing jobs and skills, organizations are taking opportunities to explore new styles to ascertaining what, talents. Hence AI casts talent management in a new role. It transforms manpower planning process from a "process developer" into an "experience architect, rethinking every facet of the HR organization.

REFERENCES

Bal, Y., Bozkurt, S., & Ertemsir, E. (2012). *The Importance of Using Human Resources Information Systems (HRIS) and a Research on Determining the Success of HRIS*. Retrieved from https://www.issbs.si/press/ISBN/978-961-6813-10- 5/papers/ML12_029.pdf

Bawany, S. (2018). What you need to lead in the Industry 4.0. *Human Resources Online*. Retrieved from https://www.humanresourcesonline.net/what-you-need-to-lead-in-theindustry-4-0

Berg, A., Buffie, E. F., & Zanna, L.-F. 2018. *Robots, growth, and inequality: Should we fear the robot revolution? (The correct answer is yes)*. IMF Working Paper No. 18/116. Washington, DC: International Monetary Fund.

Biswas, S. (2019). How to Develop a Winning Skills Strategy in the Fourth Industrial Revolution. *HR Technologist*. Retrieved from https://www.hrtechnologist.com/articles/digital-transformation/how-to-develop-awinning-skills-strategy-in-the-fourth-industrial-revolution

Blass, E. (2007). *Talent Management: Maximising talent for business performance. Executive Summary November 2007 1-12*. Chartered Management Institute, Ashbridge Consulting.

CIPD. (2006b). Talent Management: Understanding the Dimensions. London: Chartered Institute of Personnel and Development (CIPD).

CIPD. (2008). *Talent Management: an overview* (July 2008 ed.). Chartered Institute of Personnel and Development.

Davenport, T., Harris, J., & Shapiro, J. (2010). Competing on Talent Analytics. *Harvard Business Review*, *88*(10), 53–58.

Dery, K., Grant, D., Harley, B., & Wright, C. (2006a). Work, organisation and Enterprise Resource Planning systems: An alternative research agenda. *New Technology, Work and Employment*, *21*(3), 199–214.

Gakovic, A., & Yardley, K. (2007). Global Talent Management at HSBC. *Organization Development Journal*, *25*(2), 201–205.

Glen & Clayton. (2006). Key Skills Retention and Motivation: The War for Talent Still Rages and Retention is the High Ground. *Industrial and Commercial Training*, *38*(1), 37-45.

Guer, Solari, & Luca. (2012). Talent Management Practices in Italy – Implications for Human Resource Development. *International Journal of Human Resource Development, 15*(1), 25-41.

Hashani, M., & Bajrami, R. (2015). *Roli i burimeve njerëzore në rritjen e investimeve të NVM-ve*. Retrieved from https://dspace.aabedu.net/bitstream/handle/123456789/324/01-Medain-Hashani-Roberta-Bajrami-Rolii-burimeve-njer%C3%ABzore-n%C3%AB-rritjen-e.pdf?sequence=1&isAllowed=y

Tubey, R., Rotich, K. J., & Kurgat, A. (2015). History, Evolution and Development of Human Resource Management: A Contemporary Perspective. *European Journal of Business and Management, 7*(9), 139–148. http://www.eajournals.org/wp-content/uploads/History-Evolution-and-Developmentof-Human-Resource-Management-A-Contemporary-Perspective.pdf

Wiblen, S., Grant, D., & Dery, K. (2010). Transitioning to a New HRIS: The Reshaping of Human Resources and Information Technology Talent. *Journal of Electronic Commerce Research, 11*(4), 251–267.

Williams, H. (2009). Job Analysis and HR Planning. In M. Thite & M. J. Kavanagh (Eds.), *Human Resource Information Systems. Basics, Applications, and Future Directions 251-276.* SAGE Publications Inc.

Wilson, P. (2010). people@work: The Future of Work and the Changing Workplace: Challenges and Issues for Australian HR Practitioners. Melbourne: Australian Human Resources Institute.

Related Readings

To continue IGI Global's long-standing tradition of advancing innovation through emerging research, please find below a compiled list of recommended IGI Global book chapters and journal articles in the areas of personnel management, human resources management, and contemporary organizations. These related readings will provide additional information and guidance to further enrich your knowledge and assist you with your own research.

Altındağ, E. (2016). Current Approaches in Change Management. In A. Goksoy (Ed.), *Organizational Change Management Strategies in Modern Business* (pp. 24–51). Hershey, PA: IGI Global. doi:10.4018/978-1-4666-9533-7.ch002

Amoako, G. K., Adjaison, G. K., & Osei-Bonsu, N. (2016). Role of Strategic Change Management in Emerging Markets: Ghanaian Perspective. In A. Goksoy (Ed.), *Organizational Change Management Strategies in Modern Business* (pp. 328–351). Hershey, PA: IGI Global. doi:10.4018/978-1-4666-9533-7.ch016

Baniata, B. A., & Alryalat, H. (2017). The Effect of Strategic Orientations Factors to Achieving Sustainable Competitive Advantage. *International Journal of E-Entrepreneurship and Innovation*, 7(1), 1–15. doi:10.4018/IJEEI.2017010101

Barbour, J. B., Gill, R., & Barge, J. K. (2018). Exploring the Intersections of Individual and Collective Communication Design: A Research Agenda. In P. Salem & E. Timmerman (Eds.), *Transformative Practice and Research in Organizational Communication* (pp. 89–108). Hershey, PA: IGI Global. doi:10.4018/978-1-5225-2823-4.ch006

Barge, J. K. (2018). Making the Case for Academic and Social Impact in Organizational Communication Research. In P. Salem & E. Timmerman (Eds.), *Transformative Practice and Research in Organizational Communication* (pp. 235–253). Hershey, PA: IGI Global. doi:10.4018/978-1-5225-2823-4.ch014

Barron, I., & Novak, D. A. (2017). i-Leadership: Leadership Learning in the Millennial Generation. In P. Ordoñez de Pablos, & R. Tennyson (Eds.), Handbook of Research on Human Resources Strategies for the New Millennial Workforce (pp. 231-257). Hershey, PA: IGI Global. doi:10.4018/978-1-5225-0948-6.ch011

Basu, K. (2017). Change Management and Leadership: An Overview of the Healthcare Industry. In P. Ordoñez de Pablos & R. Tennyson (Eds.), *Handbook of Research on Human Resources Strategies for the New Millennial Workforce* (pp. 47–64). Hershey, PA: IGI Global. doi:10.4018/978-1-5225-0948-6.ch003

Belhaj, R., & Tkiouat, M. (2015). Including Client Opinion and Employee Engagement in the Strategic Human Resource Management: An Advanced SWOT- FUZZY Decision Making Tool. *International Journal of Human Capital and Information Technology Professionals*, 6(3), 20–33. doi:10.4018/IJHCITP.2015070102

Blithe, S. J., & Wolfe, A. W. (2018). Expanding Organizational Research Methods: Analyzing Ruptures in Qualitative Research. In P. Salem & E. Timmerman (Eds.), *Transformative Practice and Research in Organizational Communication* (pp. 168–183). Hershey, PA: IGI Global. doi:10.4018/978-1-5225-2823-4.ch010

Blomme, R. J., & Lub, X. D. (2017). Routines as a Perspective for HR-Professionals: Diversity as a Driver for Routines. In P. Ordoñez de Pablos & R. Tennyson (Eds.), *Handbook of Research on Human Resources Strategies for the New Millennial Workforce* (pp. 337–350). Hershey, PA: IGI Global. doi:10.4018/978-1-5225-0948-6.ch017

Blomme, R. J., & Morsch, J. (2016). Organizations as Social Networks: The Role of the Compliance Officer as Agent of Change in Implementing Rules and Codes of Conduct. In A. Goksoy (Ed.), *Organizational Change Management Strategies in Modern Business* (pp. 110–121). Hershey, PA: IGI Global. doi:10.4018/978-1-4666-9533-7.ch006

Bronzetti, G., Baldini, M. A., & Sicoli, G. (2017). Intellectual Capital Report in the Healthcare Sector: An Analysis of a Case Study. In P. Ordoñez de Pablos & R. Tennyson (Eds.), *Handbook of Research on Human Resources Strategies for the New Millennial Workforce* (pp. 272–285). Hershey, PA: IGI Global. doi:10.4018/978-1-5225-0948-6.ch013

Bronzetti, G., Baldini, M. A., & Sicoli, G. (2017). The Measurement of Human Capital in Family Firms. In P. Ordoñez de Pablos & R. Tennyson (Eds.), *Handbook of Research on Human Resources Strategies for the New Millennial Workforce* (pp. 371–392). Hershey, PA: IGI Global. doi:10.4018/978-1-5225-0948-6.ch019

Byrd, M. Y., & Hughes, C. (2015). A Paradigm Shift for Diversity Management: From Promoting Business Opportunity to Optimizing Lived Career Work Experiences. In C. Hughes (Ed.), *Impact of Diversity on Organization and Career Development* (pp. 28–53). Hershey, PA: IGI Global. doi:10.4018/978-1-4666-7324-3.ch002

Chakraborty, M., & Wang, J. (2015). The Postmodern Approach to Career Counseling for Contemporary Organizations. In C. Hughes (Ed.), *Impact of Diversity on Organization and Career Development* (pp. 252–274). Hershey, PA: IGI Global. doi:10.4018/978-1-4666-7324-3.ch010

Charlier, S. D., Burke-Smalley, L. A., & Fisher, S. L. (2018). Undergraduate Programs in the U.S: A Contextual and Content-Based Analysis. In J. Mendy (Ed.), *Teaching Human Resources and Organizational Behavior at the College Level* (pp. 26–57). Hershey, PA: IGI Global. doi:10.4018/978-1-5225-2820-3.ch002

Cheuk, B., & McKenzie, J. (2018). Developing the Practice of Online Leadership: Lessons From the Field. In D. Kolbaek (Ed.), *Online Collaboration and Communication in Contemporary Organizations* (pp. 235–255). Hershey, PA: IGI Global. doi:10.4018/978-1-5225-4094-6.ch013

Choi, Y. (2017). Human Resource Management and Security Policy Compliance. *International Journal of Human Capital and Information Technology Professionals*, 8(3), 68–81. doi:10.4018/IJHCITP.2017070105

Clack, L. A. (2018). Teaching Teamwork in University Settings. In J. Mendy (Ed.), *Teaching Human Resources and Organizational Behavior at the College Level* (pp. 196–210). Hershey, PA: IGI Global. doi:10.4018/978-1-5225-2820-3.ch007

Cline, B. J. (2018). Using the Classical Rhetorical Concept of "Ethos" to Solve Online Collaboration Problems of Trust and Presence: The Case of Slack. In D. Kolbaek (Ed.), *Online Collaboration and Communication in Contemporary Organizations* (pp. 82–98). Hershey, PA: IGI Global. doi:10.4018/978-1-5225-4094-6.ch005

Çolakoğlu, S., Chung, Y., & Tarhan, A. B. (2016). Strategic Human Resource Management in Facilitating Organizational Change. In A. Goksoy (Ed.), *Organizational Change Management Strategies in Modern Business* (pp. 172–192). Hershey, PA: IGI Global. doi:10.4018/978-1-4666-9533-7.ch009

Dalkir, K. (2018). Challenges in Online Collaboration: The Role of Shared Vision, Trust and Leadership Style. In D. Kolbaek (Ed.), *Online Collaboration and Communication in Contemporary Organizations* (pp. 118–138). Hershey, PA: IGI Global. doi:10.4018/978-1-5225-4094-6.ch007

Dawson, V. R. (2018). Organizing, Organizations, and the Role of Social Media Conversations. In P. Salem & E. Timmerman (Eds.), *Transformative Practice and Research in Organizational Communication* (pp. 62–78). Hershey, PA: IGI Global. doi:10.4018/978-1-5225-2823-4.ch004

De Ruiter, M., Blomme, R. J., & Schalk, R. (2016). Reducing the Negative Effects of Psychological Contract Breach during Management-Imposed Change: A Trickle-Down Model of Management Practices. In A. Goksoy (Ed.), *Organizational Change Management Strategies in Modern Business* (pp. 122–142). Hershey, PA: IGI Global. doi:10.4018/978-1-4666-9533-7.ch007

Dedousis, E., & Rutter, R. N. (2016). Workforce Localisation and Change Management: The View from the Gulf. In A. Goksoy (Ed.), *Organizational Change Management Strategies in Modern Business* (pp. 301–327). Hershey, PA: IGI Global. doi:10.4018/978-1-4666-9533-7.ch015

Densten, I. (2018). Creating University Spaces of Inspiration: Examining the Critical Link Between Leading and Lecturing. In J. Mendy (Ed.), *Teaching Human Resources and Organizational Behavior at the College Level* (pp. 59–101). Hershey, PA: IGI Global. doi:10.4018/978-1-5225-2820-3.ch003

Duran, A., & Lopez, D. (2015). Women from Diverse Backgrounds in the Science, Technology, Engineering, and Math (STEM) Professions: Retention and Career Development. In C. Hughes (Ed.), *Impact of Diversity on Organization and Career Development* (pp. 214–251). Hershey, PA: IGI Global. doi:10.4018/978-1-4666-7324-3.ch009

Durst, S., & Aggestam, L. (2017). Using IT-Supported Knowledge Repositories for Succession Planning in SMEs: How to Deal with Knowledge Loss? In P. Ordoñez de Pablos & R. Tennyson (Eds.), *Handbook of Research on Human Resources Strategies for the New Millennial Workforce* (pp. 393–406). Hershey, PA: IGI Global. doi:10.4018/978-1-5225-0948-6.ch020

Efeoğlu, E. I., & Ozcan, S. (2017). The Relationship Between Social Problem Solving Ability and Burnout Level: A Field Study Among Health Professionals. In B. Christiansen & H. Chandan (Eds.), *Handbook of Research on Human Factors in Contemporary Workforce Development* (pp. 268–282). Hershey, PA: IGI Global. doi:10.4018/978-1-5225-2568-4.ch012

Elouadi, S., & Ben Noamene, T. (2017). Does Employee Ownership Reduce the Intention to Leave? In P. Ordoñez de Pablos & R. Tennyson (Eds.), *Handbook of Research on Human Resources Strategies for the New Millennial Workforce* (pp. 111–127). Hershey, PA: IGI Global. doi:10.4018/978-1-5225-0948-6.ch006

Erne, R. (2016). Change Management Revised. In A. Goksoy (Ed.), *Organizational Change Management Strategies in Modern Business* (pp. 1–23). Hershey, PA: IGI Global. doi:10.4018/978-1-4666-9533-7.ch001

Eryılmaz, M. E., & Eryılmaz, F. (2016). Change Emphasis in Mission and Vision Statements of the First 1000 Turkish Organizations: A Content Analysis. In A. Goksoy (Ed.), *Organizational Change Management Strategies in Modern Business* (pp. 352–362). Hershey, PA: IGI Global. doi:10.4018/978-1-4666-9533-7.ch017

Fogsgaard, M., Elmholdt, C., & Lindekilde, R. (2018). Power in Online Leadership. In D. Kolbaek (Ed.), *Online Collaboration and Communication in Contemporary Organizations* (pp. 139–159). Hershey, PA: IGI Global. doi:10.4018/978-1-5225-4094-6.ch008

Francisco, R., Klein, A. Z., Engeström, Y., & Sannino, A. (2018). Knowledge on the Move: Expansive Learning Among Mobile Workers. In D. Kolbaek (Ed.), *Online Collaboration and Communication in Contemporary Organizations* (pp. 179–200). Hershey, PA: IGI Global. doi:10.4018/978-1-5225-4094-6.ch010

Galli, B. J. (2018). Overlaying Human Resources Principles to the Goal: A Research Note. *International Journal of Applied Logistics*, 8(1), 20–34. doi:10.4018/IJAL.2018010102

Galli, B. J. (2018). The Lessons of Human Resource in The Theory of Constraints. *International Journal of Organizational and Collective Intelligence*, 8(1), 13–27. doi:10.4018/IJOCI.2018010102

Gedro, J. (2016). The Academic Workplace: HRD's Potential for Creating and Maintaining a Positive Organizational Culture and Climate during Organizational Change. In C. Hughes & M. Gosney (Eds.), *Bridging the Scholar-Practitioner Gap in Human Resources Development* (pp. 166–180). Hershey, PA: IGI Global. doi:10.4018/978-1-4666-9998-4.ch009

Giannouli, V. (2017). Emotional Aspects of Leadership in the Modern Workplace. In B. Christiansen & H. Chandan (Eds.), *Handbook of Research on Human Factors in Contemporary Workforce Development* (pp. 24–59). Hershey, PA: IGI Global. doi:10.4018/978-1-5225-2568-4.ch002

Giousmpasoglou, C., & Marinakou, E. (2017). Culture and Managers in a Globalised World. In P. Ordoñez de Pablos & R. Tennyson (Eds.), *Handbook of Research on Human Resources Strategies for the New Millennial Workforce* (pp. 1–27). Hershey, PA: IGI Global. doi:10.4018/978-1-5225-0948-6.ch001

Gosney, M. W. (2016). The Interplay between Theory and Practice in HRD: A Philosophical Examination. In C. Hughes & M. Gosney (Eds.), *Bridging the Scholar-Practitioner Gap in Human Resources Development* (pp. 47–65). Hershey, PA: IGI Global. doi:10.4018/978-1-4666-9998-4.ch003

Hack-Polay, D. (2018). Putting Across Tangibility: Effectiveness of Case-Study-Based Teaching of Organisational Behaviour. In J. Mendy (Ed.), *Teaching Human Resources and Organizational Behavior at the College Level* (pp. 211–225). Hershey, PA: IGI Global. doi:10.4018/978-1-5225-2820-3.ch008

Hamlin, R. G. (2016). Evidence-Based Organizational Change and Development: Role of Professional Partnership and Replication Research. In C. Hughes & M. Gosney (Eds.), *Bridging the Scholar-Practitioner Gap in Human Resources Development* (pp. 120–142). Hershey, PA: IGI Global. doi:10.4018/978-1-4666-9998-4.ch007

Hanchey, J. N. (2018). Reworking Resistance: A Postcolonial Perspective on International NGOs. In P. Salem & E. Timmerman (Eds.), *Transformative Practice and Research in Organizational Communication* (pp. 274–291). Hershey, PA: IGI Global. doi:10.4018/978-1-5225-2823-4.ch016

Hassan, A., & Rahimi, R. (2017). Insights and Rumination of Human Resource Management Practices in SMEs: Case of a Family Run Tour Operator in London. In P. Ordoñez de Pablos & R. Tennyson (Eds.), *Handbook of Research on Human Resources Strategies for the New Millennial Workforce* (pp. 258–271). Hershey, PA: IGI Global. doi:10.4018/978-1-5225-0948-6.ch012

Herd, A., & Alagaraja, M. (2016). Strategic Human Resource Development Alignment: Conceptualization from the Employee's Perspective. In C. Hughes & M. Gosney (Eds.), *Bridging the Scholar-Practitioner Gap in Human Resources Development* (pp. 85–100). Hershey, PA: IGI Global. doi:10.4018/978-1-4666-9998-4.ch005

Hieker, C., & Rushby, M. (2017). Diversity in the Workplace: How to Achieve Gender Diversity in the Workplace. In B. Christiansen & H. Chandan (Eds.), *Handbook of Research on Human Factors in Contemporary Workforce Development* (pp. 308–332). Hershey, PA: IGI Global. doi:10.4018/978-1-5225-2568-4.ch014

Huffman, T. (2018). Imagination, Action, and Justice: Trends and Possibilities at the Intersection of Organizational Communication and Social Justice. In P. Salem & E. Timmerman (Eds.), *Transformative Practice and Research in Organizational Communication* (pp. 292–306). Hershey, PA: IGI Global. doi:10.4018/978-1-5225-2823-4.ch017

Hughes, C. (2015). Integrating Diversity into Organization and Career Development: A Changing Perspective. In C. Hughes (Ed.), *Impact of Diversity on Organization and Career Development* (pp. 1–27). Hershey, PA: IGI Global. doi:10.4018/978-1-4666-7324-3.ch001

Hughes, C. (2015). Leveraging Diversity for Competitive Advantage. In C. Hughes (Ed.), *Impact of Diversity on Organization and Career Development* (pp. 275–298). Hershey, PA: IGI Global. doi:10.4018/978-1-4666-7324-3.ch011

Hughes, C., & Gosney, M. W. (2016). Human Resource Development as a Knowledge Management System: The Importance of Bridging the Scholar-Practitioner Gap. In C. Hughes & M. Gosney (Eds.), *Bridging the Scholar-Practitioner Gap in Human Resources Development* (pp. 1–19). Hershey, PA: IGI Global. doi:10.4018/978-1-4666-9998-4.ch001

Hughes, C., & Stephens, D. (2016). Use Value and HRD and HRM Flexibility: Implications for HRD Practice. In C. Hughes & M. Gosney (Eds.), *Bridging the Scholar-Practitioner Gap in Human Resources Development* (pp. 181–199). Hershey, PA: IGI Global. doi:10.4018/978-1-4666-9998-4.ch010

Jahn, J. (2018). Doing Applied Organizational Communication Research: Bridging a Gap Between Our and Managers' Understandings of Organization and Communication. In P. Salem & E. Timmerman (Eds.), *Transformative Practice and Research in Organizational Communication* (pp. 221–234). Hershey, PA: IGI Global. doi:10.4018/978-1-5225-2823-4.ch013

Jensen, I. (2018). Online Leadership and Communication Across Cultures: Developing an Interdisciplinary Approach. In D. Kolbaek (Ed.), *Online Collaboration and Communication in Contemporary Organizations* (pp. 64–81). Hershey, PA: IGI Global. doi:10.4018/978-1-5225-4094-6.ch004

Jeong, S., Lim, D. H., & Park, S. (2017). Leadership Convergence and Divergence in the Era of Globalization. In P. Ordoñez de Pablos & R. Tennyson (Eds.), *Handbook of Research on Human Resources Strategies for the New Millennial Workforce* (pp. 286–309). Hershey, PA: IGI Global. doi:10.4018/978-1-5225-0948-6.ch014

Jha, J. K., & Singh, M. (2017). Human Resource Planning as a Strategic Function: Biases in Forecasting Judgement. *International Journal of Strategic Decision Sciences*, 8(3), 120–131. doi:10.4018/IJSDS.2017070106

King, D. R. (2016). Management as a Limit to Organizational Change: Implications for Acquisitions. In A. Goksoy (Ed.), *Organizational Change Management Strategies in Modern Business* (pp. 52–73). Hershey, PA: IGI Global. doi:10.4018/978-1-4666-9533-7.ch003

Kishna, T., Blomme, R. J., & van der Veen, J. A. (2016). Organizational Routines: Developing a Duality Model to Explain the Effects of Strategic Change Initiatives. In A. Goksoy (Ed.), *Organizational Change Management Strategies in Modern Business* (pp. 363–385). Hershey, PA: IGI Global. doi:10.4018/978-1-4666-9533-7.ch018

Kolbaek, D. (2018). Design-Based Research as a Methodology for Studying Learning in the Context of Work: Suggestions for Guidelines. In D. Kolbaek (Ed.), *Online Collaboration and Communication in Contemporary Organizations* (pp. 21–42). Hershey, PA: IGI Global. doi:10.4018/978-1-5225-4094-6.ch002

Kolbaek, D. (2018). Online Leaders Increase Three Types of Capital. In D. Kolbaek (Ed.), *Online Collaboration and Communication in Contemporary Organizations* (pp. 99–117). Hershey, PA: IGI Global. doi:10.4018/978-1-5225-4094-6.ch006

Kolbaek, D. (2018). Online Leadership and Learning: How Online Leaders May Learn From Their Working Experience. In D. Kolbaek (Ed.), *Online Collaboration and Communication in Contemporary Organizations* (pp. 201–219). Hershey, PA: IGI Global. doi:10.4018/978-1-5225-4094-6.ch011

Konyu-Fogel, G. (2015). Career Management and Human Resource Development of a Global, Diverse Workforce. In C. Hughes (Ed.), *Impact of Diversity on Organization and Career Development* (pp. 80–104). Hershey, PA: IGI Global. doi:10.4018/978-1-4666-7324-3.ch004

Kuhn, T. (2018). Working and Organizing as Social Problems: Reconceptualizing Organizational Communication's Domain. In P. Salem & E. Timmerman (Eds.), *Transformative Practice and Research in Organizational Communication* (pp. 30–42). Hershey, PA: IGI Global. doi:10.4018/978-1-5225-2823-4.ch002

Kyeyune, C. N. (2016). Career Development Models and Human Resource Development Practice. In C. Hughes & M. Gosney (Eds.), *Bridging the Scholar-Practitioner Gap in Human Resources Development* (pp. 66–84). Hershey, PA: IGI Global. doi:10.4018/978-1-4666-9998-4.ch004

Lakshminarayanan, S. (2016). Corporate Trainers: Practitioner-Scholars in the Workplace. In C. Hughes & M. Gosney (Eds.), *Bridging the Scholar-Practitioner Gap in Human Resources Development* (pp. 143–165). Hershey, PA: IGI Global. doi:10.4018/978-1-4666-9998-4.ch008

Leonardi, P. (2018). The Process of Theorizing in Organizational Communication: On the Importance of Owning Phenomena. In P. Salem & E. Timmerman (Eds.), *Transformative Practice and Research in Organizational Communication* (pp. 80–88). Hershey, PA: IGI Global. doi:10.4018/978-1-5225-2823-4.ch005

Maheshkar, C. (2016). HRD 'Scholar-Practitioner': An Approach to Filling Theory, Practice and Research Gap. In C. Hughes & M. Gosney (Eds.), *Bridging the Scholar-Practitioner Gap in Human Resources Development* (pp. 20–46). Hershey, PA: IGI Global. doi:10.4018/978-1-4666-9998-4.ch002

Malik, A. (2016). The Role of HR Strategies in Change. In A. Goksoy (Ed.), *Organizational Change Management Strategies in Modern Business* (pp. 193–215). Hershey, PA: IGI Global. doi:10.4018/978-1-4666-9533-7.ch010

Marinakou, E., & Giousmpasoglou, C. (2017). Gendered Leadership as a Key to Business Success: Evidence from the Middle East. In P. Ordóñez de Pablos & R. Tennyson (Eds.), *Handbook of Research on Human Resources Strategies for the New Millennial Workforce* (pp. 200–230). Hershey, PA: IGI Global. doi:10.4018/978-1-5225-0948-6.ch010

Martins, A., Martins, I., & Pereira, O. (2017). Challenges Enhancing Social and Organizational Performance. In P. Ordóñez de Pablos & R. Tennyson (Eds.), *Handbook of Research on Human Resources Strategies for the New Millennial Workforce* (pp. 28–46). Hershey, PA: IGI Global. doi:10.4018/978-1-5225-0948-6.ch002

Martins, A., Martins, I., & Pereira, O. (2017). Embracing Innovation and Creativity through the Capacity of Unlearning. In P. Ordóñez de Pablos & R. Tennyson (Eds.), *Handbook of Research on Human Resources Strategies for the New Millennial Workforce* (pp. 128–147). Hershey, PA: IGI Global. doi:10.4018/978-1-5225-0948-6.ch007

Matuska, E. M., & Grubicka, J. (2017). Employer Branding and Internet Security. In B. Christiansen & H. Chandan (Eds.), *Handbook of Research on Human Factors in Contemporary Workforce Development* (pp. 357–378). Hershey, PA: IGI Global. doi:10.4018/978-1-5225-2568-4.ch016

Meisenbach, R. J. (2018). Ethics, Agency, and Non-Human Agency in the Study of the Communicative Constitution of Organizations. In P. Salem & E. Timmerman (Eds.), *Transformative Practice and Research in Organizational Communication* (pp. 255–273). Hershey, PA: IGI Global. doi:10.4018/978-1-5225-2823-4.ch015

Mendy, J. (2018). Key HRM Challenges and Benefits: The Contributions of the HR Scaffolding. In J. Mendy (Ed.), *Teaching Human Resources and Organizational Behavior at the College Level* (pp. 1–24). Hershey, PA: IGI Global. doi:10.4018/978-1-5225-2820-3.ch001

Mendy, J. (2018). Rethinking the Contribution of Organizational Change to the Teaching and Learning of Organizational Behaviour and Human Resource Management: The Quest for Balance. In J. Mendy (Ed.), *Teaching Human Resources and Organizational Behavior at the College Level* (pp. 103–132). Hershey, PA: IGI Global. doi:10.4018/978-1-5225-2820-3.ch004

Mukhopadhyay, P. (2017). Investigation of Ergonomic Risk Factors in Snacks Manufacturing in Central India: Ergonomics in Unorganized Sector. In B. Christiansen & H. Chandan (Eds.), *Handbook of Research on Human Factors in Contemporary Workforce Development* (pp. 425–449). Hershey, PA: IGI Global. doi:10.4018/978-1-5225-2568-4.ch019

Mulhall, S., & Campbell, M. (2018). Embedding Career Competencies in Learning and Talent Development: Career Management and Professional Development Modules. In J. Mendy (Ed.), *Teaching Human Resources and Organizational Behavior at the College Level* (pp. 133–171). Hershey, PA: IGI Global. doi:10.4018/978-1-5225-2820-3.ch005

Muralidharan, E., & Pathak, S. (2017). National Ethical Institutions and Social Entrepreneurship. In B. Christiansen & H. Chandan (Eds.), *Handbook of Research on Human Factors in Contemporary Workforce Development* (pp. 379–402). Hershey, PA: IGI Global. doi:10.4018/978-1-5225-2568-4.ch017

Naik, K. R., & Srinivasan, S. R. (2017). Distinctive Leadership: Moral Identity as Self Identity. In P. Ordoñez de Pablos & R. Tennyson (Eds.), *Handbook of Research on Human Resources Strategies for the New Millennial Workforce* (pp. 90–110). Hershey, PA: IGI Global. doi:10.4018/978-1-5225-0948-6.ch005

Naito, Y. (2017). Factors Related to Readjustment to Daily Life: A Study of Repatriates in Japanese Multinational Enterprises. In B. Christiansen & H. Chandan (Eds.), *Handbook of Research on Human Factors in Contemporary Workforce Development* (pp. 403–424). Hershey, PA: IGI Global. doi:10.4018/978-1-5225-2568-4.ch018

Nawaz, T. (2017). Expatriation in the Age of Austerity: An Analysis of Capital Mobilization Strategies of Self-Initiated Expatriates. In P. Ordoñez de Pablos & R. Tennyson (Eds.), *Handbook of Research on Human Resources Strategies for the New Millennial Workforce* (pp. 177–199). Hershey, PA: IGI Global. doi:10.4018/978-1-5225-0948-6.ch009

Nielsen, A. S. (2018). Professional Collaboration in a World Without Offices: The Case of a Co-Working Space on Bali. In D. Kolbaek (Ed.), *Online Collaboration and Communication in Contemporary Organizations* (pp. 276–291). Hershey, PA: IGI Global. doi:10.4018/978-1-5225-4094-6.ch015

Nortvig, A. (2018). Technology and Presence: Multi-Presence in Online Interactions. In D. Kolbaek (Ed.), *Online Collaboration and Communication in Contemporary Organizations* (pp. 220–233). Hershey, PA: IGI Global. doi:10.4018/978-1-5225-4094-6.ch012

Özgeldi, M. (2016). Role of Human Resources in Change. In A. Goksoy (Ed.), *Organizational Change Management Strategies in Modern Business* (pp. 216–229). Hershey, PA: IGI Global. doi:10.4018/978-1-4666-9533-7.ch011

Palm, K. (2017). A Case of Phased Retirement in Sweden. In P. Ordoñez de Pablos & R. Tennyson (Eds.), *Handbook of Research on Human Resources Strategies for the New Millennial Workforce* (pp. 351–370). Hershey, PA: IGI Global. doi:10.4018/978-1-5225-0948-6.ch018

Patro, C. S. (2017). Performance Appraisal System Effectiveness: A Conceptual Review. In B. Christiansen & H. Chandan (Eds.), *Handbook of Research on Human Factors in Contemporary Workforce Development* (pp. 156–180). Hershey, PA: IGI Global. doi:10.4018/978-1-5225-2568-4.ch007

Patro, C. S. (2017). Welfare Regime: A Critical Discourse. In B. Christiansen & H. Chandan (Eds.), *Handbook of Research on Human Factors in Contemporary Workforce Development* (pp. 110–131). Hershey, PA: IGI Global. doi:10.4018/978-1-5225-2568-4.ch005

Pietiläinen, V., Salmi, I., Rusko, R., & Jänkälä, R. (2017). Experienced Stress and the Value of Rest Stops in the Transportation Field: Stress and Transportation. In B. Christiansen & H. Chandan (Eds.), *Handbook of Research on Human Factors in Contemporary Workforce Development* (pp. 249–267). Hershey, PA: IGI Global. doi:10.4018/978-1-5225-2568-4.ch011

Pilny, A. N., & Poole, M. S. (2018). An Introduction to Computational Social Science for Organizational Communication. In P. Salem & E. Timmerman (Eds.), *Transformative Practice and Research in Organizational Communication* (pp. 184–200). Hershey, PA: IGI Global. doi:10.4018/978-1-5225-2823-4.ch011

Pries-Heje, J., & Pries-Heje, L. (2018). Improving Online Collaboration in Contemporary IT Development Teams. In D. Kolbaek (Ed.), *Online Collaboration and Communication in Contemporary Organizations* (pp. 160–178). Hershey, PA: IGI Global. doi:10.4018/978-1-5225-4094-6.ch009

Richet, J. (2016). Internal Communication Failure in Times of Change. In A. Goksoy (Ed.), *Organizational Change Management Strategies in Modern Business* (pp. 289–300). Hershey, PA: IGI Global. doi:10.4018/978-1-4666-9533-7.ch014

Riemann, U. (2016). The Power of Three: A Blended Approach of Project-, Change Management, and Design Thinking. In A. Goksoy (Ed.), *Organizational Change Management Strategies in Modern Business* (pp. 74–94). Hershey, PA: IGI Global. doi:10.4018/978-1-4666-9533-7.ch004

Roach, C. M., & Davis-Cooper, G. (2016). An Evaluation of the Adoption of the Integrated Human Resource Information System in Trinidad and Tobago. *International Journal of Public Administration in the Digital Age*, *3*(3), 1–17. doi:10.4018/IJPADA.2016070101

Salem, P. J. (2018). Transformative Organizational Communication Practices. In P. Salem & E. Timmerman (Eds.), *Transformative Practice and Research in Organizational Communication* (pp. 109–129). Hershey, PA: IGI Global. doi:10.4018/978-1-5225-2823-4.ch007

Salem, P. J., & Timmerman, C. E. (2018). Forty Years of Organizational Communication. In P. Salem & E. Timmerman (Eds.), *Transformative Practice and Research in Organizational Communication* (pp. 1–28). Hershey, PA: IGI Global. doi:10.4018/978-1-5225-2823-4.ch001

Schramm, M. (2018). The Virtual Coffee Break: Virtual Leadership – How to Create Trust and Relations Over Long Distances. In D. Kolbaek (Ed.), *Online Collaboration and Communication in Contemporary Organizations* (pp. 256–275). Hershey, PA: IGI Global. doi:10.4018/978-1-5225-4094-6.ch014

Scott, C. L., & Sims, J. D. (2015). Workforce Diversity Career Development: A Missing Piece of the Curriculum in Academia. In C. Hughes (Ed.), *Impact of Diversity on Organization and Career Development* (pp. 129–150). Hershey, PA: IGI Global. doi:10.4018/978-1-4666-7324-3.ch006

Scott, C. R., & Kang, K. K. (2018). Invisible Domains and Unexplored Terrains: A Multi-Level View of (In)Appropriately Hidden Organizations. In P. Salem & E. Timmerman (Eds.), *Transformative Practice and Research in Organizational Communication* (pp. 43–61). Hershey, PA: IGI Global. doi:10.4018/978-1-5225-2823-4.ch003

Seino, K., Nomoto, A., Takezawa, T., & Boeltzig-Brown, H. (2017). The Diversity Management for Employment of the Persons With Disabilities: Evidence of Vocational Rehabilitation in the United States and Japan. In B. Christiansen & H. Chandan (Eds.), *Handbook of Research on Human Factors in Contemporary Workforce Development* (pp. 333–356). Hershey, PA: IGI Global. doi:10.4018/978-1-5225-2568-4.ch015

Shen, L., & Austin, L. (2017). Communication and Job Satisfaction. In B. Christiansen & H. Chandan (Eds.), *Handbook of Research on Human Factors in Contemporary Workforce Development* (pp. 201–225). Hershey, PA: IGI Global. doi:10.4018/978-1-5225-2568-4.ch009

Sims, C. H. (2015). Genderized Workplace Lookism in the U.S. and Abroad: Implications for Organization and Career Development Professionals. In C. Hughes (Ed.), *Impact of Diversity on Organization and Career Development* (pp. 105–127). Hershey, PA: IGI Global. doi:10.4018/978-1-4666-7324-3.ch005

Şimşit, Z. T., Günay, N. S., & Vayvay, Ö. (2016). Organizational Learning to Managing Change: Key Player of Continuous Improvement in the 21st Century. In A. Goksoy (Ed.), *Organizational Change Management Strategies in Modern Business* (pp. 95–109). Hershey, PA: IGI Global. doi:10.4018/978-1-4666-9533-7.ch005

Sklaveniti, C. (2018). Theorizing Virtual Teams: Relationality in Dispersed Collaboration. In D. Kolbaek (Ed.), *Online Collaboration and Communication in Contemporary Organizations* (pp. 1–20). Hershey, PA: IGI Global. doi:10.4018/978-1-5225-4094-6.ch001

Starr-Glass, D. (2017). The Misappropriation of Organizational Power and Control: Managerial Bullying in the Workplace. In B. Christiansen & H. Chandan (Eds.), *Handbook of Research on Human Factors in Contemporary Workforce Development* (pp. 87–109). Hershey, PA: IGI Global. doi:10.4018/978-1-5225-2568-4.ch004

Stephens, K. K. (2018). Adapting and Advancing Organizational Communication Research Methods: Balancing Methodological Diversity and Depth, While Creating Methodological Curiosity. In P. Salem & E. Timmerman (Eds.), *Transformative Practice and Research in Organizational Communication* (pp. 151–167). Hershey, PA: IGI Global. doi:10.4018/978-1-5225-2823-4.ch009

Teimouri, H., Jenab, K., Moazeni, H. R., & Bakhtiari, B. (2017). Studying Effectiveness of Human Resource Management Actions and Organizational Agility: Resource Management Actions and Organizational Agility. *Information Resources Management Journal, 30*(2), 61–77. doi:10.4018/IRMJ.2017040104

Thoms, C. L., & Burton, S. L. (2015). Understanding the Impact of Inclusion in Disability Studies Education. In C. Hughes (Ed.), *Impact of Diversity on Organization and Career Development* (pp. 186–213). Hershey, PA: IGI Global. doi:10.4018/978-1-4666-7324-3.ch008

Tkachenko, O., Hahn, H., & Peterson, S. (2016). Theorizing the Research-Practice Gap in the Field of Management: A Review of Key Frameworks and Models. In C. Hughes & M. Gosney (Eds.), *Bridging the Scholar-Practitioner Gap in Human Resources Development* (pp. 101–119). Hershey, PA: IGI Global. doi:10.4018/978-1-4666-9998-4.ch006

Tomasiak, M. A., & Chamakiotis, P. (2017). Understanding Diversity in Virtual Work Environments: A Comparative Case Study. In B. Christiansen & H. Chandan (Eds.), *Handbook of Research on Human Factors in Contemporary Workforce Development* (pp. 283–307). Hershey, PA: IGI Global. doi:10.4018/978-1-5225-2568-4.ch013

Torlak, N. G. (2016). Improving the Role of Organisational Culture in Change Management through a Systems Approach. In A. Goksoy (Ed.), *Organizational Change Management Strategies in Modern Business* (pp. 230–271). Hershey, PA: IGI Global. doi:10.4018/978-1-4666-9533-7.ch012

Tracy, S. J., & Donovan, M. C. (2018). Moving From Practical Application to Expert Craft Practice in Organizational Communication: A Review of the Past and OPPT-ing Into the Future. In P. Salem & E. Timmerman (Eds.), *Transformative Practice and Research in Organizational Communication* (pp. 202–220). Hershey, PA: IGI Global. doi:10.4018/978-1-5225-2823-4.ch012

Tran, B. (2017). The Art and Science in Communication: Workplace (Cross-Cultural) Communication Skills and Competencies in the Modern Workforce. In B. Christiansen & H. Chandan (Eds.), *Handbook of Research on Human Factors in Contemporary Workforce Development* (pp. 60–86). Hershey, PA: IGI Global. doi:10.4018/978-1-5225-2568-4.ch003

Treem, J. W., & Barley, W. C. (2018). A Framework for How Expertise Is Communicated and Valued in Contemporary Organizations: Why Process Work Matters. In P. Salem & E. Timmerman (Eds.), *Transformative Practice and Research in Organizational Communication* (pp. 130–149). Hershey, PA: IGI Global. doi:10.4018/978-1-5225-2823-4.ch008

Trinh, M. P. (2015). When Demographic and Personality Diversity are Both at Play: Effects on Team Performance and Implications for Diversity Management Practices. In C. Hughes (Ed.), *Impact of Diversity on Organization and Career Development* (pp. 54–79). Hershey, PA: IGI Global. doi:10.4018/978-1-4666-7324-3.ch003

Trusson, C. (2018). The Call to Teach Human Capital Analytics. In J. Mendy (Ed.), *Teaching Human Resources and Organizational Behavior at the College Level* (pp. 173–195). Hershey, PA: IGI Global. doi:10.4018/978-1-5225-2820-3.ch006

Umamaheswari, S., & Krishnan, J. (2017). Retention Factor: Work Life Balance and Policies – Effects over Different Category of Employees in Ceramic Manufacturing Industries. In P. Ordoñez de Pablos & R. Tennyson (Eds.), *Handbook of Research on Human Resources Strategies for the New Millennial Workforce* (pp. 329–336). Hershey, PA: IGI Global. doi:10.4018/978-1-5225-0948-6.ch016

Wahyuningtyas, R., & Anggadwita, G. (2017). Perspective of Managing Talent in Indonesia: Reality and Strategy. In P. Ordoñez de Pablos & R. Tennyson (Eds.), *Handbook of Research on Human Resources Strategies for the New Millennial Workforce* (pp. 407–420). Hershey, PA: IGI Global. doi:10.4018/978-1-5225-0948-6.ch021

Washington, G. D., & Shen, L. (2017). Emotional Intelligence and Job Stress. In B. Christiansen & H. Chandan (Eds.), *Handbook of Research on Human Factors in Contemporary Workforce Development* (pp. 226–248). Hershey, PA: IGI Global. doi:10.4018/978-1-5225-2568-4.ch010

Wichmand, M. (2018). Is Urgent Evoke a Digital Ba?: How a Game Can Make Space for Knowledge Creation. In D. Kolbaek (Ed.), *Online Collaboration and Communication in Contemporary Organizations* (pp. 44–63). Hershey, PA: IGI Global. doi:10.4018/978-1-5225-4094-6.ch003

Wittmer, J. L., & Rudolph, C. W. (2015). The Impact of Diversity on Career Transitions over the Life Course. In C. Hughes (Ed.), *Impact of Diversity on Organization and Career Development* (pp. 151–185). Hershey, PA: IGI Global. doi:10.4018/978-1-4666-7324-3.ch007

Yildirim, F., Abukan, B., & Oztas, D. (2017). Determining the Needs for Employee Assistance Programs (EAPs): A Comparative Study on Public and Private Sector Employees. In P. Ordoñez de Pablos & R. Tennyson (Eds.), *Handbook of Research on Human Resources Strategies for the New Millennial Workforce* (pp. 65–89). Hershey, PA: IGI Global. doi:10.4018/978-1-5225-0948-6.ch004

You, J., Kim, J., & Lim, D. H. (2017). Organizational Learning and Change: Strategic Interventions to Deal with Resistance. In P. Ordoñez de Pablos & R. Tennyson (Eds.), *Handbook of Research on Human Resources Strategies for the New Millennial Workforce* (pp. 310–328). Hershey, PA: IGI Global. doi:10.4018/978-1-5225-0948-6.ch015

You, J., Kim, J., & Miller, S. M. (2017). Organizational Learning as a Social Process: A Social Capital and Network Approach. In B. Christiansen & H. Chandan (Eds.), *Handbook of Research on Human Factors in Contemporary Workforce Development* (pp. 132–155). Hershey, PA: IGI Global. doi:10.4018/978-1-5225-2568-4.ch006

Zel, U. (2016). Leadership in Change Management. In A. Goksoy (Ed.), *Organizational Change Management Strategies in Modern Business* (pp. 272–288). Hershey, PA: IGI Global. doi:10.4018/978-1-4666-9533-7.ch013

Zheng, W., Wu, Y. J., & Xu, M. (2017). From Democratic Participation to Shared Values: Improving Employee-Employer Interactions to Achieve Win-Win Situations. In P. Ordoñez de Pablos & R. Tennyson (Eds.), *Handbook of Research on Human Resources Strategies for the New Millennial Workforce* (pp. 421–432). Hershey, PA: IGI Global. doi:10.4018/978-1-5225-0948-6.ch022

About the Contributors

Anchal Pathak is working as an Associate Professor with Bule Hora University, Ethiopia. Having more than 7 years of teaching and research experience in the field of Human Resource Management and Organizational Behaviour with excellent communication and teaching skills.

Shikha Rana is currently associated with IMS Unison University as an Assistant Professor-Senior Scale. She has an experience of more than nine years and her area of expertise is OB/HRM. She is an assiduous researcher and possesses excellent interpersonal skills.

* * *

Preeti Bhaskar is a research scholar at ICFAI University, Dehradun and working as faculty at University of Technology and Applied Sciences, Ibra, Oman. She possesses 9 years of teaching experience in the area of Human Resource Management. Her research interest includes Technology adoption, E-government, Job performance, Job Satisfaction, sustainable development, continuing education and E-learning. She has published research papers in many reputed journals (ABDC & SCOPUS) and presented research papers at various national and international conferences. She has also authored two books on "general Management" & published two case studies in Case Centre, the United Kingdom. She has also completed two minor research projects sponsored by the Symbiosis international university, Pune. She is actively engaged in conducting student development programs and faculty development programs at various colleges and universities.

Raluca Bunduchi is Senior Lecturer in Innovation at the University of Edinburgh Business School. Her research lies at the intersection of innovation management, information systems and organisational research. Her current work explores how actors engage in and make sense of different forms of innovation, ranging from the introduction of new kinds of digital process technology in established organisations, to the development and commercialisation of new forms of social products and services in new ventures. She has published in a range of journals including The Journal of Product Innovation Management, Technovation, Information Systems Journal and Information & Management.

Richa Das is an Assistant professor at Atria Institute of Technology. Her field of specialization is in Human resource Management and Organisational Behaviour. She has more than 8 years of research and teaching experience. She has completed her Ph.D from IIT (ISM), Dhanbad. Her area of research interests are: women empowerment and development, Human Resource Management and entrepreneurship. She has published six papers in SCOPUS and SCI indexed journal. An avid researcher and teacher, she loves digging herself into new areas of research and learn new topics.

Minisha Gupta is a Leadership Mentor in Industry & Academic Excellence. She has worked as a researcher and academician in the field of HRM for more than 5 years. Her research interests include HRM, Organizational Change, Innovation Management, Talent Management, Entrepreneurship and Intrapreneurship.

Amit Joshi is triple post-graduate in Economics, Commerce, and English; he has completed his Master's Diploma in International Trade from Symbiosis, Pune. He has been awarded a degree of M.Phil in Economics and is a Doctorate in English. Currently, he is working as the head of the department at ICFAI Business School, ICFAI University, Dehradun. His area of expertise includes English, Business Communication and Life Skills. He has more than 18 years of teaching experience and has several research publications in various reputed journals to his credit. He has also authored poetry book "Stimulus: From darkness to light (A collection of motivational poems)". He has presented research papers in several national and international conferences that include places like Dusseldorf, Germany, Saint Petersburg, Russia, and University of Dubai, etc. His research interests include communication strategy, the art of writing, disseminating knowledge, personal effectiveness, marginal productivity theory, trickledown effect and communication modeling. He has conducted many faculty development programs, student development programs, and corporate training programs.

Pushpa Kataria, an experienced trainer, career spanning more than 18 years in the field of Management and Strategy. Besides a management educationist, she is a socialist and corporate trainer. She has initiated, designed and conducted numerous Management Development Programs & In-Company training programs at cross functional managerial levels in private and public sector organizations. She has presented and published research papers & book chapters in International and

National Journals of repute. She is also a National Scholarship Holder from Govt. of Andhra Pradesh at school level. She holds an enriched academic career.

Sanjeev Kumar is an Associate Professor in the School of Hotel Management at Lovely Professional University, Punjab. He has more than 17 years of experience. He has written many research papers on Human Resource, food and service domains. He has attended many conferences and seminars.

Claudia Pagliari is an interdisciplinary researcher specialising in the study of digital innovations in the contexts of healthcare, business, government, and society. Undertakes empirical and analytical research, including evaluation of interventions and programmes, evidence synthesis, and policy studies. Interested in the ethics and governance of digital and data-driven initiatives.

Teena Saharan, a proficient trainer and researcher, has an extensive work experience of 16 years. She is an expert in conducting experiential learning based trainings & development programs in area of Women and Executive Leadership Development, Emotional Intelligence and Effective communication etc. She has authored two books on Industrial Psychology and Industrial Sociology and presented and published many research papers in places of International Repute. She is a certified trainer in areas of 'HR metrics and Analytics' and 'Psychometric Testing'.

Amrik Singh is an Associate Professor in the School of Hotel Management at Lovely Professional University, Punjab. He has written many research papers on Human Resource, Accommodation management and Hospitality domains. He also serving as reviewer in many Scopus indexed journals and also written many papers for Scopus indexed journals.

Vandana Singh is an Assistant Professor in School of Management at IMS Unison University. She is Ph.D., UGC (NET/JRF), MBA, and has a teaching experience of 1.5 years. She has to her credit 12 publications in referred journals which includes one listed in SCOPUS. She has participated in various workshops based on SPSS, Intellectual Rights and Research Methodology and has presented research papers in various International and National Conferences & Seminars. She has served as a reviewer for reputed journals published by IGI-Global.

Aizhan Tursunbayeva works as an Assistant Manager at KPMG Advisory S.p.A., and tutors the Managing Change course at the University of Edinburgh and Human Resources course at the University of Molise. Her research lies at the intersection of HRM, information systems and healthcare. She has published in a range of journals including Information Technology & People, Journal of the American Medical Informatics Association, and Management Learning.

Muddu Vinay is the Dean, IBS & Vice Chancellor at the ICFAI University Dehradun. He has Vinay has around three decades of rich experience in the field of Education, Training, Planning, Management and Consultancy. He is acknowledged not only for establishing Institutions but also leading them to excellence in Academics and Research with his path breaking initiatives. Dr. Muddu Vinay has held several positions of eminence in professional bodies in the field of education both globally and in India. These include Chairman IQAC(Internal Quality Assessment Cell), Proctor, University of California, Berkeley Extension. He has spearheaded and pioneered various programmes in the field of higher education especially in postgraduate studies in collaboration with reputed schools in India, Australia, New Zealand, UK, Singapore and Malaysia. He is the present expert peer committee member, National Board of Accreditation, Govt. of India, New Delhi. He has also been the expert member for various professional organizations that include Higher Education UK India Education and Research Initiative (UKIERI), World Bank and MHRD. He has been conferred with 'Award of Excellence- Eminent Educationists' in the IndoAmerican summit 2016, MTC Global Award for Excellence 2013 in 'Innovation in Teaching Pedagogy', 'NESA Fellowship Award 2012' and 'Leadership and Responsibility Award' by Presidency Group of Institutions, along with many other awards and accolades in his illustrious career. He has personally trained more than 30,000 individuals undertaking undergraduate, post graduate studies as well as research scholars, University teachers and teaching staff. He is widely acclaimed for creating successful business models in the education sector without compromising on the ethical dimensions of teaching as a profession.

Index

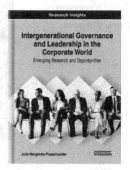

Ensure Quality Research is Introduced to the Academic Community

Become an IGI Global Reviewer for Authored Book Projects

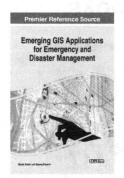
Premier Reference Source

Emerging GIS Applications for Emergency and Disaster Management

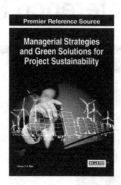
Premier Reference Source

Managerial Strategies and Green Solutions for Project Sustainability

Premier Reference Source

Comparative Approaches to Using R and Python for Statistical Data Analysis

Premier Reference Source

Solutions for High-Touch Communications in a High-Tech World

The overall success of an authored book project is dependent on quality and timely reviews.

In this competitive age of scholarly publishing, constructive and timely feedback significantly expedites the turnaround time of manuscripts from submission to acceptance, allowing the publication and discovery of forward-thinking research at a much more expeditious rate. Several IGI Global authored book projects are currently seeking highly-qualified experts in the field to fill vacancies on their respective editorial review boards:

Applications and Inquiries may be sent to:
development@igi-global.com

Applicants must have a doctorate (or an equivalent degree) as well as publishing and reviewing experience. Reviewers are asked to complete the open-ended evaluation questions with as much detail as possible in a timely, collegial, and constructive manner. All reviewers' tenures run for one-year terms on the editorial review boards and are expected to complete at least three reviews per term. Upon successful completion of this term, reviewers can be considered for an additional term.

If you have a colleague that may be interested in this opportunity, we encourage you to share this information with them.

IGI Global Proudly Partners With eContent Pro International

Receive a 25% Discount on all Editorial Services

Editorial Services

IGI Global expects all final manuscripts submitted for publication to be in their final form. This means they must be reviewed, revised, and professionally copy edited prior to their final submission. Not only does this support with accelerating the publication process, but it also ensures that the highest quality scholarly work can be disseminated.

English Language Copy Editing

Let eContent Pro International's expert copy editors perform edits on your manuscript to resolve spelling, punctuaion, grammar, syntax, flow, formatting issues and more.

Scientific and Scholarly Editing

Allow colleagues in your research area to examine the content of your manuscript and provide you with valuable feedback and suggestions before submission.

Figure, Table, Chart & Equation Conversions

Do you have poor quality figures? Do you need visual elements in your manuscript created or converted? A design expert can help!

Translation

Need your documjent translated into English? eContent Pro International's expert translators are fluent in English and more than 40 different languages.

Email: customerservice@econtentpro.com   www.igi-global.com/editorial-service-partners

Printed in the United States
By Bookmasters